8
0-33894

Martin B-57 Canberra

Robert C. Mikesh

U.S. AIR FORCE
033879
B
33908

Martin B-57 Canberra

The Complete Record

Robert C. Mikesh

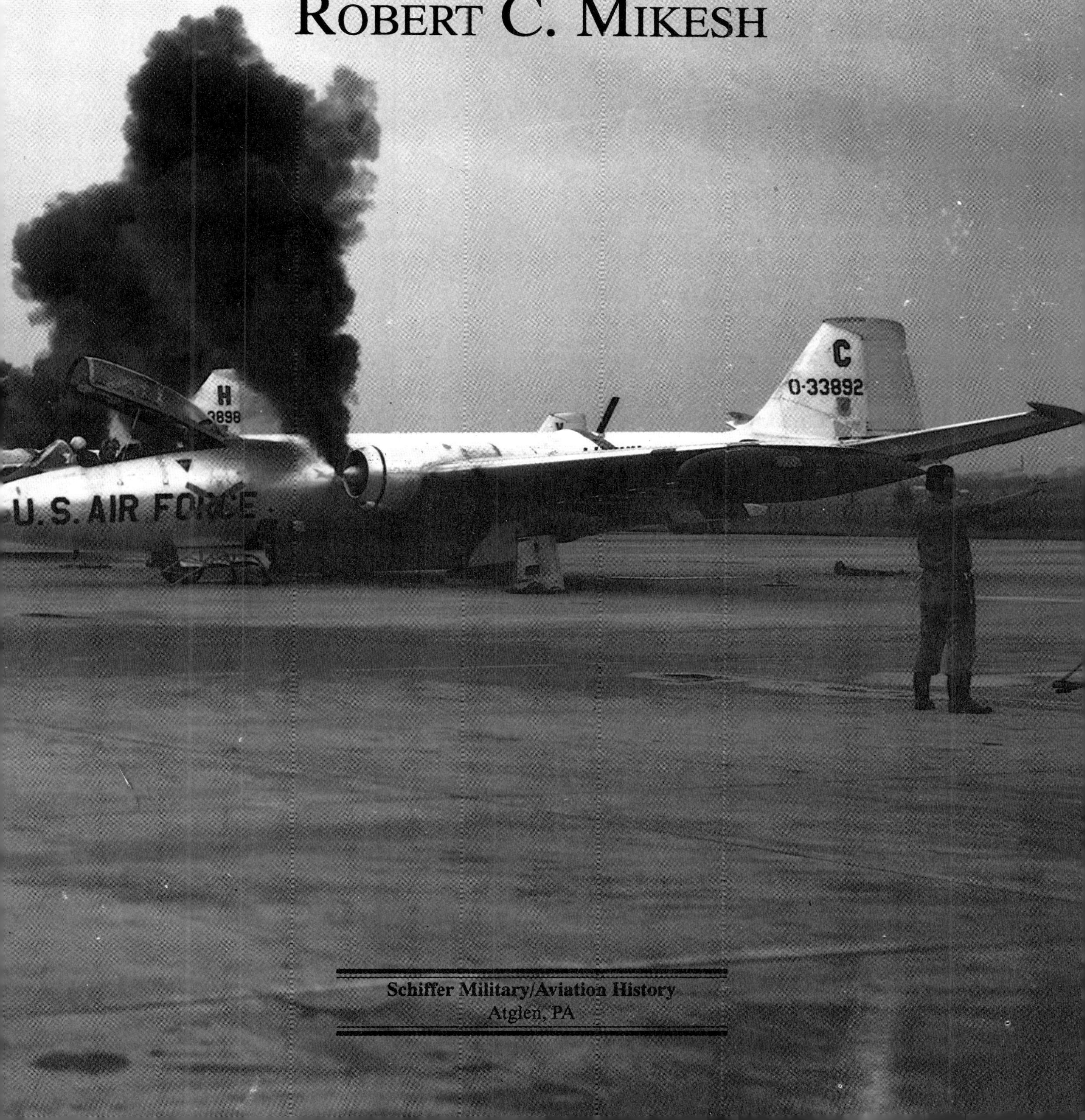

Schiffer Military/Aviation History
Atglen, PA

Acknowledgements

Space does not permit the names of all who have contributed in some degree to this history about the Canberra. Nor is there space to mention the specialty area in which a person was particularly helpful. My gratitude to all must be accepted by this sincere thanks for their frequent, timely, and willing help. Heading this list – and rightfully so, is Ramona, my patient wife, who occupied herself during times we would otherwise share together, while I was absorbed over the long time period of writing this book. She was always willing to proof read and offer constructive suggestions when needed. Without her encouragement and faithful help, this book would not have been written.

Helping immeasurably with the British development portion of this coverage was Roland P. Beamont, English Electric's test pilot for the Canberra, and G.M. Hobday, English Electric resident advisory engineer to the Glenn L. Martin Company during B-57 production. They have provided the balance needed in comparing the British to the American-built aircraft.

Information in many forms that range from personal accounts, documentation, to photographs have come from many people, all of which is so necessary to piece together this coverage of the B-57 throughout its many years of service. To achieve this, I have relied heavily upon the help received from David A. Anderton, Douglas J. Beggerly, Dana Bell, George R. Boucher, Ellis E. Bruch, Frank R. Chandler, Robert P. Coombs, Peter Cotellesse, Richard A. Crist, Robert A. Danielson, Larry Davis, Carlton 'Jake' Dolwick, Robert F. Dorr, Gene R. Durden, Melvin C. Eisaman, Judy Endicott, Steve Feaster, Donald E. Felts, Keith Ferris, Wm. A. Fleming, James L. Goodnight, Lloyd J. Goodrow, Irwin 'Pat' Graham, Paul R. Guthals, Joseph A. Guthrie, Jr, Roger D. Harrington, Gerard Hasselwander, Kenneth H. High, Robert Hunter, Lynn M. John, Albert T. Keeler, John N. Kendrick, Charles D. Leonard, John W. Little, Robert Lince, Robert D. Lutes, Rex Marlowe, Larry B. Mason, Walter J. McGinnis, Roger T. McLeod (RAAF), David W. Menard, Royce F. Miller, Clyde E. Millington, Dudley L. Mizer, Lee A. Mongeon, Donald N. Nation, Stephane Nicolaou, Paul R. Pitt, David T. Reynolds, Gerry I. Repenon, Jerry W. Russell, James J. Scanlon, David Schoem, Robert A. Schultz, J.T. Stanley, Howard A. Steady, Peter Steinemann (RAAF), Charles Strain, E. 'Ted' Sturmthal, O.E. 'Pat' Tibbs, Norman Thompson, Gerry Turner (RAF), Darrell Van Citters, Robin A. Walker, and Eugene E. Youngs. To all those who have helped, I convey my grateful thanks.

Robert C. Mikesh
Washington, D.C. 1994

An abridged edition of this book was published in 1980 by Ian Allan Ltd.

Book Design by Robert Biondi.

Library of Congress Catalog Number: 94-66964

Printed in China.
ISBN: 0-88740-661-0

We are interested in hearing from authors with book ideas on related topics.

Published by Schiffer Publishing Ltd.
77 Lower Valley Road
Atglen, PA 19310
Please write for a free catalog.
This book may be purchased from the publisher.
Please include $2.95 postage.
Try your bookstore first.

Barry Goldwater had been Senator from Arizona for 30 years and once was a candidate for President of the United States. He is a staunch advocate for a strong military force for the United States and had served many years on the Senate Armed Services Committee. In this capacity and as an Air Force Reserve Major-General, having received his military wings during World War II, he took every opportunity to acquaint himself, firsthand, with the flying equipment of the military services. In pursuit of this, Senator Goldwater has several hundred hours, first pilot time, in the B-57 Canberra.

FOREWORD

A new generation of aircraft and pilots has come into the US Air Force since the B-57 was phased out of military service. In fact, when mentioning to present day pilots that the B-57 was an airplane that I had once flown, there is question in my mind if they are certain what I am talking about. With the phase-out of the B-57 Canberra in 1983, so went the era of the tactical bomber that had its beginning before World War II. In fact, while the B-57 was in Air Force service, it was the only tactical bomber in inventory. Replacing its mission is what we know today as 'strike aircraft' ranging from fighter-bombers to the 'A' series of aircraft like the A-10.

The B-57 did not pass into oblivion easily. In fact, in some circles it was considered an 'old' airplane when it flew combat in Vietnam, yet they stayed around for nearly another twenty years after dropping their first bombs in anger.

My experience as a B-57 pilot began with 'Star Flight' at Andrews AFB, near Washington, D.C., in the early 1960s. This unit was comprised of a handful of B-57s and pilots set aside for administrative flights for those of us who could share the flying duties as well. It was here that I first met author Robert Mikesh and we often flew together when I had to quickly visit a distant Air Force Base. Straight and level flying is fine for getting somewhere, but there were opportunities to see what the airplane could really do. After Bob demonstrated a loop off the deck in this bird, then allowed me to try my hand, I couldn't help but exclaim, 'By golly, this really flies like an airplane should.' I was impressed with its maneuverability which was far better than I would have expected of an aircraft of this size. Its soundness and flexibility in mission profile made it a superb combat weapon, and it served in that capacity year after year in Southeast Asia compiling a remarkable record for itself. It is amazing, considering the length of time this aircraft was in service, that it performed so well in so many different capacities. Because of this, it gathered hundreds of admirers over the years, especially from those of us who have had the rewarding firsthand experience of flying and working with this rugged, versatile airplane – the B-57 Canberra.

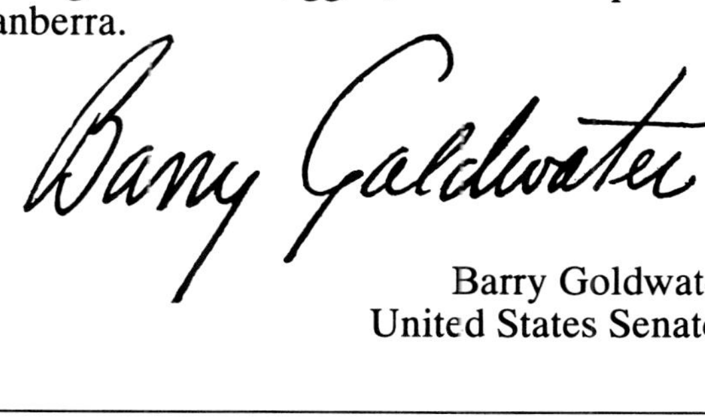

Barry Goldwater
United States Senator

Contents

INTRODUCTION

My first glimpse of the Canberra was the classic photo of it in a steeply banked turn at the introductory show at Farnborough in September 1949. My reaction after disbelief, was relief – that this was not a USAF plane that I might one day have to fly. To me, its straight, wide and stubby wings looked totally antiquated. How quickly we change, for in a short time I was waiting anxiously for our unit to be equipped with the American-built version of this bomber, and from the first time I flew the B-57B, I was hooked on it forever.

Time has proven that the straightforward design was correct, for the basic Canberra airframe design stayed in continuous U.S. military service until 1983, and in other countries even longer. As the immortal DC-3 is the great workhorse of the air transports, the Canberra certainly occupies a similar niche in history among combat aircraft.

This acceptance by the United States Air Force in lieu of American types is a lasting tribute to the British design team that created it. By the time production B-57s were reaching tactical bomber units, the Douglas B-66, also in production, was being converted to other missions even before its acceptance in the tactical bomber role. For years the B-57 remained the only jet tactical bomber in the USAF. When it had dropped its last bomb, there were no pure tactical bombers to replace it nor has there been since.

Perhaps the term 'tactical bomber' in itself is obsolete, for 'attack aircraft' and 'fighter-bombers' fill this mission requirement of close air support that the B-57 did so well. The term as well as B-57s themselves, nearly came to an end as far back as the late 1950s when tactical bomber units began phasing out at a steady rate due to mandatory military cutbacks. Only one combat wing remained (yet it too was scheduled for deactivation) when an escalating war situation in Southeast Asia reversed Air Force thinking. The Canberra played an important part in the nine years of combat that followed. When the fighting ended, the B-57s were removed from the role of USAF combat aircraft, and the dwindling numbers retained in service were reassigned to a peacetime mission. At this writing however, one WB-57F still remains in service for high altitude evaluation missions for NACA.

Earlier than I like to remember, I decided one day to write a book about the B-57, for I was in my third pilot assignment with Canberras (later to become five B-57 unit assignments, fifteen years, and 2,000 hours) and they had become an important part of my life. I began gathering information about the airplane and its crews and soon had more material about the B-57 than one book could handle. That book became *B-57 Canberra At War 1964-1972* published by Ian Allan in 1980. In time, that book became out of print, depriving this story of the B-57 from others that were interested or became interested as a new generation of readers. Fortunately, Schiffer Publishing Ltd. agreed to reissue the book and in so doing allowed me the opportunity to insert many details that had to be left out of the original, make adjustments where needed, and to add more to the story that had surfaced from the first edition. We are fortunate to have color photographs added to this volume.

The overall story of the American-built Canberra is contained in both books, but readers close to the subject will recognize that some aspects of its history have been passed over lightly, and others omitted completely. One book cannot contain all the detail I would like – the total story of the B-57, an airplane that served in so many capacities and stayed around as a U.S. military airplane for 29 years.

Much of the credit for the content of this book is shared with many people, including fellow Canberra pilots who had some facts more clearly in their mind than I. Many of their names appear throughout this history, for airplanes and people cannot be separated. The B-57 has touched many people – those that designed and built the Canberra, the crews that flew it, and those that supported it so caringly on the ground – I only regret that everyone's contributions cannot be recognized within this history of an aircraft that will not be easily forgotten.

1

BIRTH OF THE CANBERRA

The merits of an airplane can often be measured against what it was designed to do and how well it meets those requirements. The success of the Canberra is not only attributed to having fulfilled the initial design requirements, but due to the 'stretch' capability of its basic design, it became highly adaptable to a multitude of other tasks as well.

To trace the concept of the Canberra to its very earliest glimmer may well have been the idea that Mr. W.E.W. Petter had for a jet fighter-bomber replacement for the Royal Air Force's Whirlwinds and Typhoons. When 'Teddy' Petter was Technical Director for Westland Aircraft Ltd in 1944, the jet age had already began, and any tactical airplane having a propeller was immediately declared obsolete. It was here that Petter's jet aircraft concepts began developing before he left Westland. Not only did Petter have a single engine concept along the lines illustrated here, but his thinking began with a twin-engine concept with both engines in the fuselage.

It was that same year, 1944, that Petter moved to Preston, Lancashire. He become Chief Designer for English Electric Company Ltd, leading a new design team, for English Electric had not built an airplane of their own design since 1926. The company had suspended its aviation activities that year and this had lasted until 1938 when the war years brought English Electric back into the

First flown on 13 May 1949, the cleanly contoured English Electric Canberra was Britain's first jet bomber. Its concept was so right that it remained in production for 12 years, and some are still in service at this writing. Low wing loading and a low aspect ratio wing were designed into the Canberra, giving it fighter-like handling qualities, as well as high altitude capability resulting in maximum fuel economy.

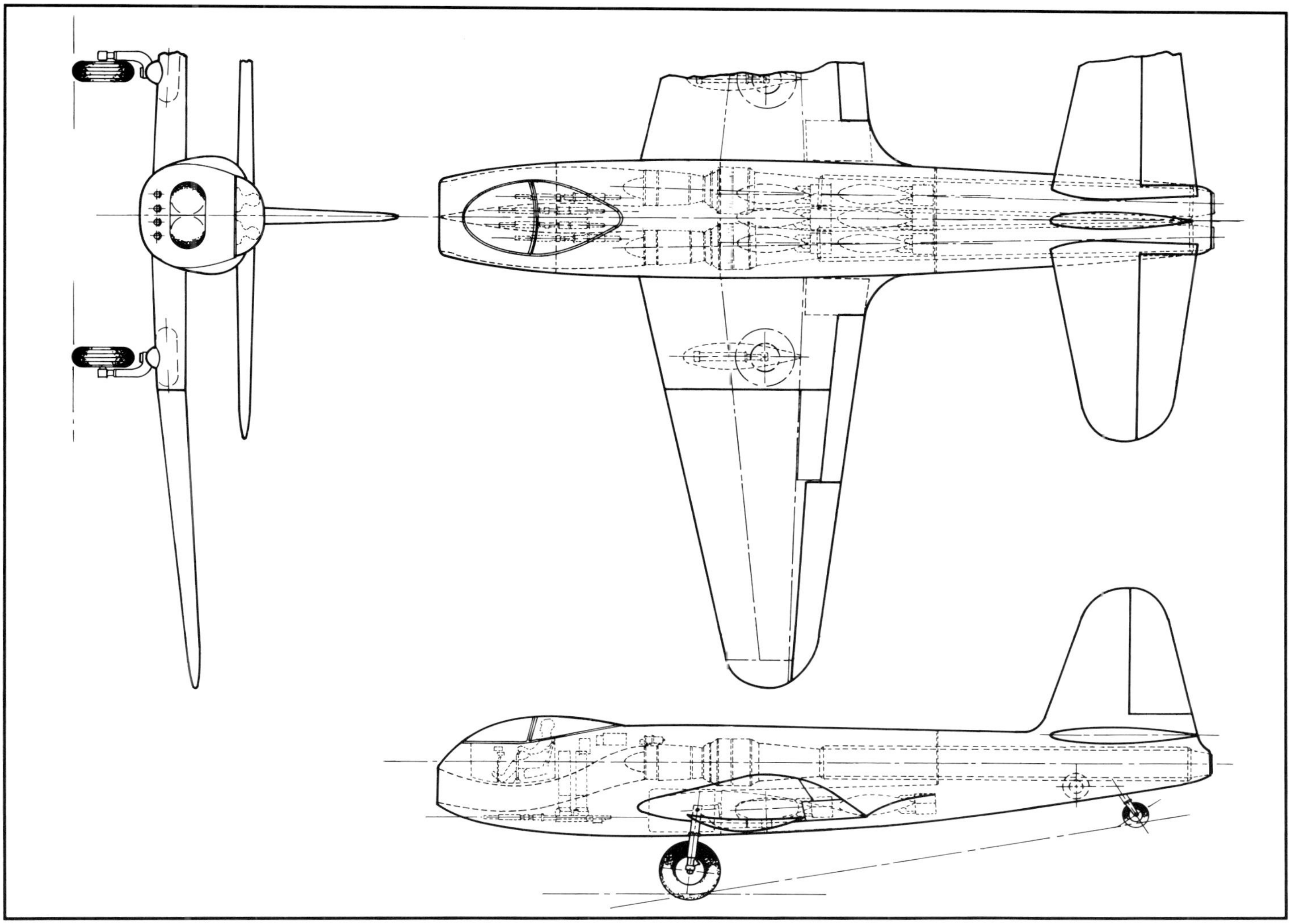

The earliest concept of a jet powered aircraft that is traceable to the Canberra is this design by W.E.W. Petter when still working at Westland in 1944. Upon leaving Westland for English Electric, this twin jet engine fighter-bomber concept became the embryo that led to the Canberra.

aviation field by manufacturing aircraft of other companies, namely the Halifax and Hampden bombers. When jet powered aircraft came into the picture, the Ministry of Aircraft Production invited English Electric to develop a jet aircraft design written around their own proposal of meeting certain broad requirements.

Starting with his earlier thoughts on jet aircraft design, Petter and his new design team members considered a jet bomber concept, one that might have the qualities that would replace all RAF bomber types then in service, considering in particular the mission requirements ranging from the Mosquito to the four-engine Lincolns.

Design work began, and by June 1945 the new airplane had the lines of a mid-wing monoplane, powered by a single, very large turbojet, crewed by a pilot and navigator. The entire design was built around a proposed Rolls-Royce engine which was to have a 5ft 6in diameter, which they planned to mount in the center of the fuselage. This two-stage centrifugal engine, though unusually large, was expected to produce 12,000lb static thrust – enough to make anyone take notice in those early jet years. The single unit concept was decided upon after a careful study had been made of other engines then available, and all were considered inadequate for a bomber. Although the craft would be large, weighing in the neighborhood of 40,000lb, it would cruise at 500mph at between 35 and 40,000ft. Even in this embryonic stage, the design showed a resemblance to the now familiar lines of the Canberra.

Before the design study became too advanced, the rapid development of jet engines led the English Electric design team to re-examine its plan, and in so doing dropped the single engine concept substituting two engines buried in the wing-roots. The new engines were axial flow Rolls-Royce AJ65s, a forerunner of the Avon, having a much smaller frontal area than the originally considered centrifugal engine. With the oversized engine no longer in the fuselage, the bomb bay area was expanded to carry an assortment of weights including one 8,000lb bomb. Fuel tankage was reduced since the axial flow engine had a lower specific fuel consumption for the designed range of the bomber. The elimination of the fuselage jet tail-pipe also made possible a simpler and lighter weight aft section. Engine air inlets remained in the leading edge of the wing which had been established in the single engine design.

When the newer Rolls-Royce engine appeared in late 1945, the design was again modified by placing these engines within the wing at about the one third span point. As weight reduction continued in the design, the wing area was able to be reduced from 1,040

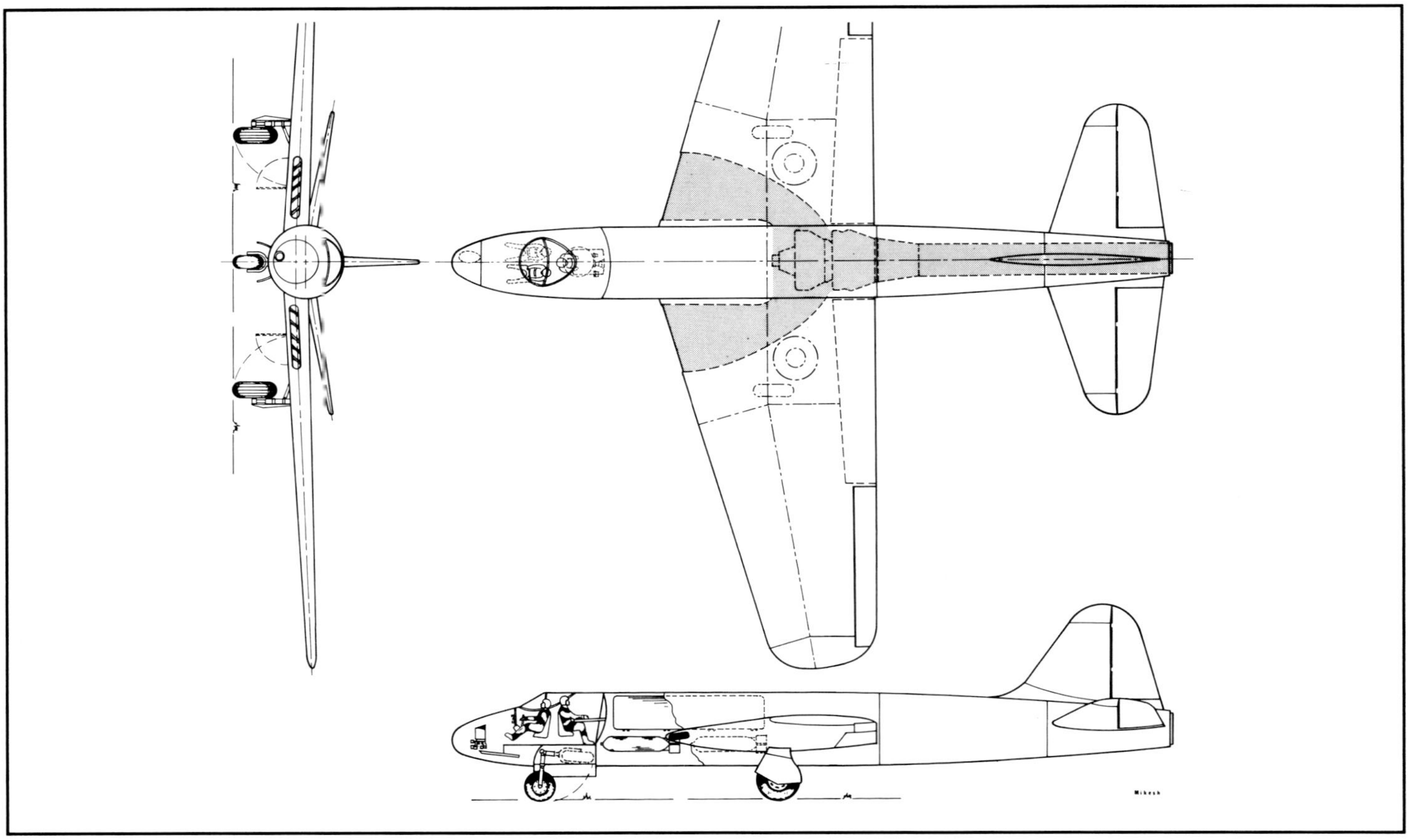
Mikesh

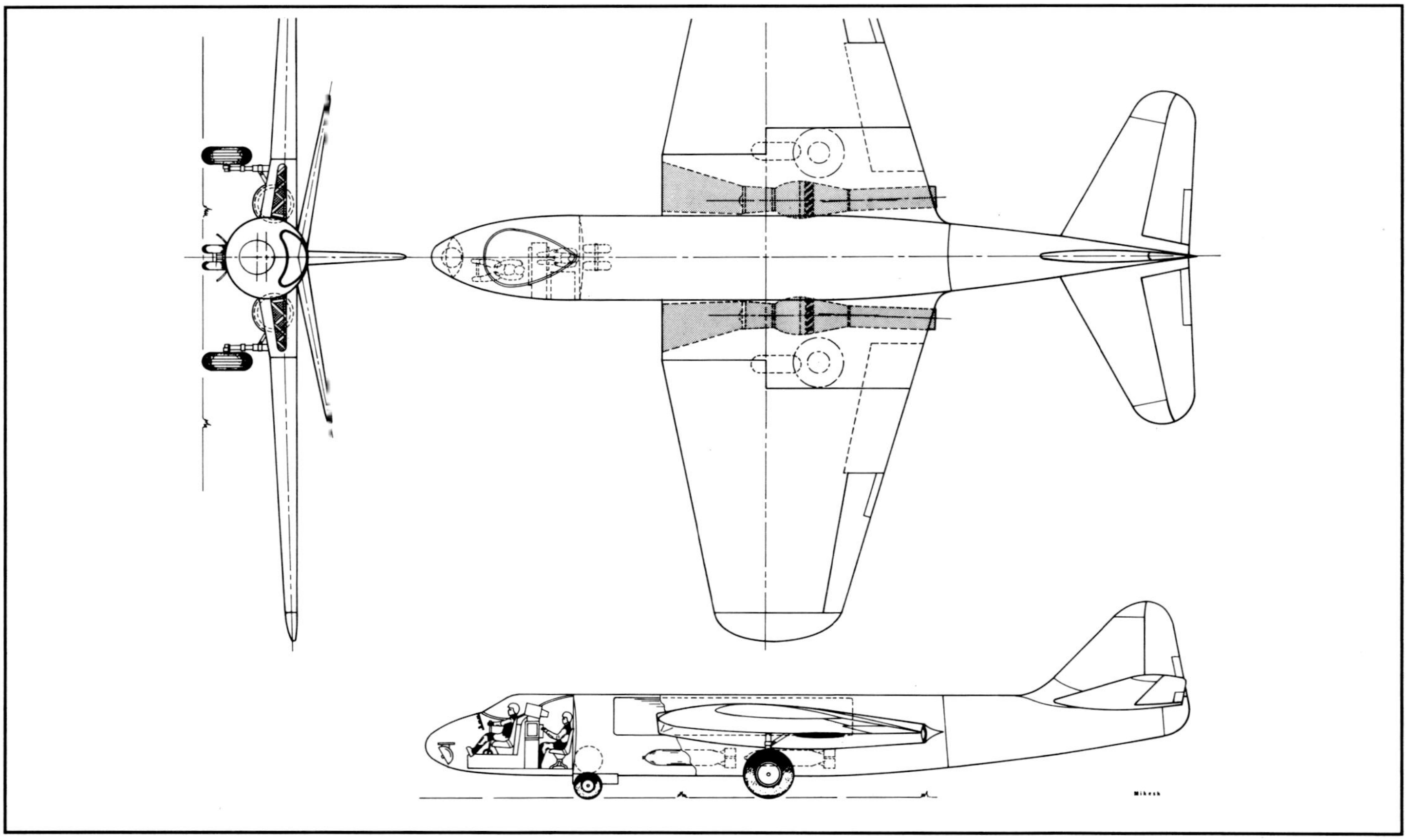
Mikesh

OPPOSITE: Early transitional designs leading to the Canberra were these two designs. In 1945 this jet bomber was a single engine concept having the centrifugal-flow jet engine buried in the fuselage. By 1946 with the advent of axial-flow jet engines, the proposal had two of these engines mounted in the wing roots. Design changes that followed, evolved in a logical development pattern that became the prototype Canberra.

to 960sq ft without increasing the planned wing loading, and the aspect ratio was also reduced from 5.4 to 4.9. With this third conceptual design for a jet bomber, the configuration we now recognize as the Canberra began to take form.

Since the newly organized design team was a new addition to the English Electric Company, their initial home was temporarily in a wartime acquired garage in Preston and it was there, in the lower level of the building, that the shape of the Canberra was put together in the form of a wooden mock-up. The design met the Ministry of Supply B3/45 specification and on 7 January 1946, a contract for four prototype English Electric A1 aircraft (as the unnamed Canberras were then called) were ordered for production and test evaluation.

A whole new field of performance factors complicated the initial design strength for this and all other jet planes. As aircraft speeds approached the compressibility stage near the speed of sound, a new set of structural strength rules had to be developed relative to the speed of sound. The measurement of speed in the conventional manner of miles-per-hour, had no relationship to speed of sound, which varies with height and temperature. This new method of speed

This may well be the first roll-out picture of the Canberra in preparation for initial engine run-up. The open side hatch is for cockpit access for the three crew members. Initial concept of the Canberra was that of a high altitude radar bomber. This accounts for the radar-dome nose later changed to clear. (BAC)

values came in the form of Mach numbers, where Mach I equalled the speed of sound, and measurement of speed is made from this point. The Canberra was not intended to be a transonic or supersonic aircraft when carrying a useful military load since it was unlikely that greater thrust would be obtainable from two Avons during the operational life of the aircraft. Instead it was to operate at high altitude in the Mach 0.7 range, therefore its design could remain uncomplicated because it introduced few engineering problems associated with speeds closer to Mach 1, which were then

Britain's first Canberra B1 nears final assembly at the former American aerodrome near Warton. It was from here that the first flight took place on May 13, 1949. The uncowled engines were a marvel for their smaller diameter than reciprocating engines of comparable power. (BAC)

The grand-daddy of them all, VN799, the B1, Britain's first jet bomber, designed and built by the English Electric Co Ltd., now British Aerospace Corporation. In its earliest form, this prototype had a stabilizing dorsal fin, and rounded rudder tip, later to be deleted. (BAC)

difficult to overcome. This simplicity was the secret of its success. It was built to meet certain performance factors in the most practical structural form, thus eliminating complications in unproven design theories for high speed.

The wings of the Canberra were not swept back as no advantage would be gained at the maximum speed for which the aircraft was designed to fly. According to 'Teddy' Petter, the technical solution was found to lie primarily in the right choice of wing. A light wing loading, a comparatively low aspect ratio, a smooth structure and a modest thickness/chord ratio were all deemed to be prerequisites. In addition, great strength could be built into the wing because of its thickness, without being a deterrent to its speed. The depth of 27in was a mere 12% in relation to its fore-and-aft measurement of 19ft which was large for an airplane of this size, and the upper and lower surface curved equally about the fore-and-aft center line of the wing. This symmetrical airfoil section was employed as a means of avoiding violent pitch-changes when the critical Mach number was exceeded, yet without sacrifice of good control characteristics.

A high altitude performance was the objective for the design of this airplane rather than speed. When measured against the Meteor, which had a practical intercept altitude of 40,000ft, the design of the Canberra with its low wing loading, was to have the same turning ability at 50,000ft, giving the bomber a drastically superior

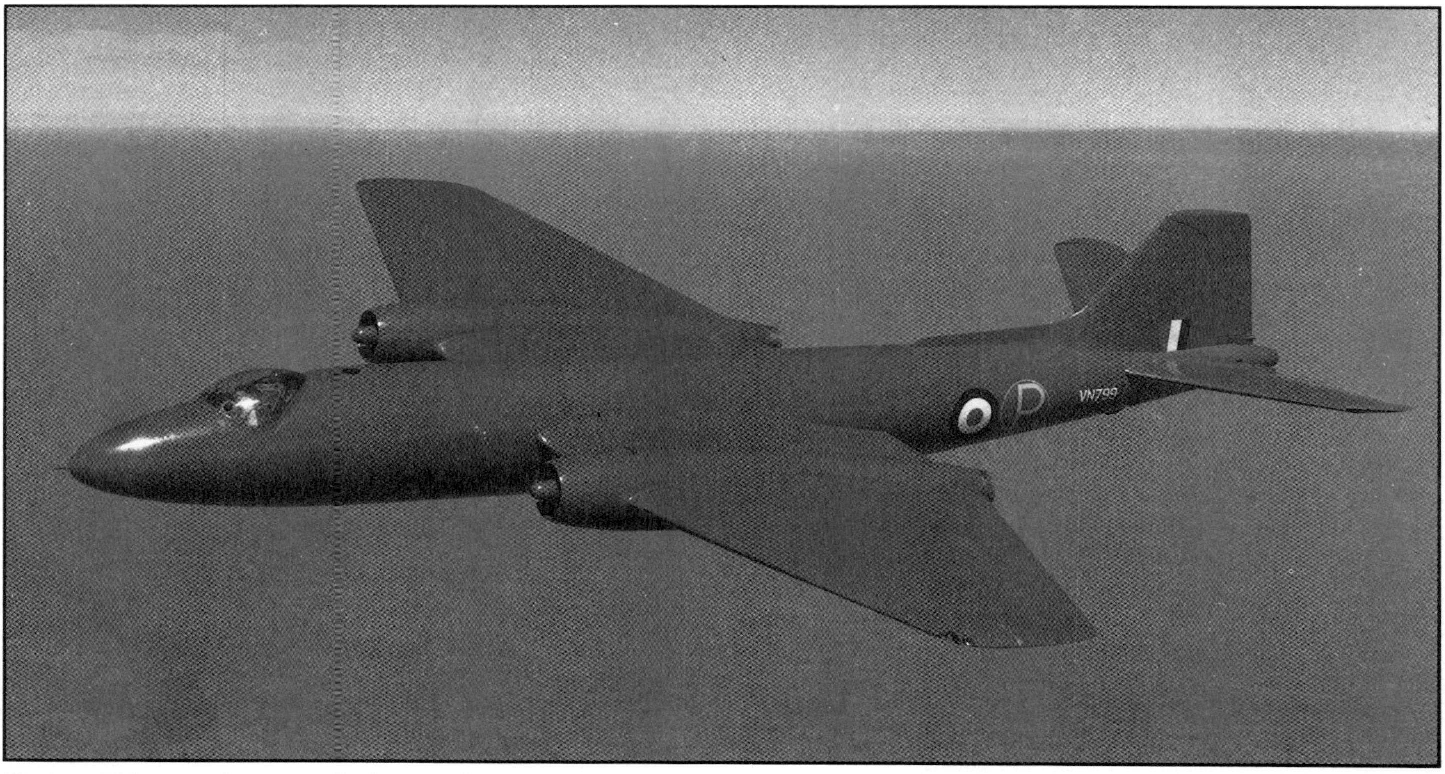

The Avon RA2 powered prototype Canberra in the air. Tip of rudder has already been trimmed which proved necessary after the first flight. Dorsal fin was eliminated on production models. Bareheaded Beamont can be seen in the cockpit of the blue painted craft. (BAC)

performance over fighters of that period. In anticipation of this mark of performance, defensive armament was not considered in the design.

As the construction of the first Canberra began in near wartime security at Preston, English Electric acquired the former American aerodrome at nearby Warton, where final assembly and first flights would take place. By early May 1949, taxi tests began with Roland Beamont at the controls. Beamont had been with the company for two years after leaving the RAF as a wing commander, and had followed the development of this jet bomber from its beginning.

Once completed, the Canberra was proven ready for flight with three 10ft high hops during taxi trials that took place on the day before the scheduled first flight. These hops were at Beamont's insistence, for this would confirm the effectiveness of flight controls before being fully committed to the air. The fact that each of these hops was for a distance of about 1,200ft after a take-off run, and then landing, all well within the runway length and without over stressing with wheel brakes, was the first practical proof of the exceptional qualities of the airplane.

When the day for the first flight arrived, Petter was overheard to ask Beamont if he 'really wanted to fly today?' It was Friday, 13 May 1949. Beamont responded, however, that 'it was as good as any other day,' and proceeded to take the Canberra off the runway at Warton for its maiden flight which lasted 27 minutes.

All that was hoped to be accomplished on that initial flight of England's first jet-bomber was realized. The only uneasiness experienced during the flight was a sharp directional jerk each time rudder pressure was applied. Once assured that rudder trim was centered, no further rudder movements were made that might aggravate the situation. Following the flight, Beamont made the following entries in the flight report which serve to illustrate his enthusiasm for the airplane:

> 'Apart from the rudder condition described, the aircraft handled smoothly and easily. All services operated satisfactorily although in the case of tailplane actuation some alterations (as to rate) may be necessary . . .
>
> 'Both engines and airframe were remarkably quiet in flight and the noise level in the cockpit allows excellent radio reception . . .
>
> 'Rudder and aileron trimmers were set at neutral for take-off and were not required again in the flight . . .'

Following the flight, and based on Beamont's recommendations, the rounded top of the rudder which included the balance horn, was trimmed down, and no further suggestion of directional problems were encountered. This accounts for the squared off appearance of the rudder that carried into production models.

On flights that followed, Beamont discovered that the aircraft performed more like a fighter than a bomber. Aerobatics had not been written into the design requirements but, flown within the design limits of speed and 'G' forces, there was nothing to limit the airplane from rolls and loops. Exploring its potential away from inquisitive eyes, Beamont prepared a flight routine for the coming Society of British Aircraft Constructors' Exhibition and Display at Farnborough that September, which would punctuate the Canberra's first public showing. As expected, the airplane stole the show. The

This is the American Martin-built B-57 which was developed from the British Canberra. There is no noticeable exterior differences. The interior layout of the British Canberra was three place with two crewmen seated side by side behind the pilot. The American B-57 had only two crew positions, the second being seated low and behind the pilot. (Martin)

American publication *Aviation Week* tells of this often talked-of event:

> 'Canberra Shows Off – biggest military surprise of the show was the English Electric Co Ltd sky-blue Canberra jet bomber. US observers were not impressed with the Canberra's straight wing and somewhat conventional configuration on the ground. But in the air the combination of test pilot R. P. Beamont and the 15,000lb thrust from the two axial Avons made the Canberra behave in spectacular fashion.
>
> 'Its speed range from 500 to less than 100 mph was ably demonstrated by Beamont who followed his high-speed passes on the deck with an approach using full flaps and gear down and bomb bay doors open that slowed the Canberra to less than 100mph. At this speed he rocked the big bomber violently with ailerons to show the full control available as it approached stalling speed.
>
> 'Beamont whipped the bomber (designed to carry a 10,000lb bomb load) around on the deck like a fighter, flying it through a series of slow rolls, high speed turns and remarkable rates of climb. The Canberra was originally designed for radar bombing at around 50,000ft, but Beamont's demonstration convinced many Britishers the new bomber may prove to be another Mosquito in its versatility at everything from low-level attack through high fighting to high altitude bombing.'*

The Canberra met with immediate acceptance. Production orders for the RAF's new bomber were filled not only by English Electric, but Handley Page Ltd., A. V. Roe, and Short Brothers & Harland as well, to a total of 546 machines. Capitalising on its speed and performance, it was quick to set innumerable records throughout the world. Its model variants are countless, as well as its service in air forces of other countries. As a combat plane for what it was designed to be, it has proven its worth in more than one conflict. The list of British Canberra's accomplishments will be endless since a number of British-built Canberras remain in service at this writing. But that is another book – and another story.

* *Aviation Week & Space Technology*

2

AMERICANS BECOME INTERESTED

American spectators at early demonstrations of the Canberra were equally as impressed as all others, but the US military observers could not envisage what they might do with the airplane if it were theirs! As one reporter put it, 'it is neither "fish nor fowl" in that it is slightly too large to be a fighter, slightly too small to even be a "light" bomber.' To Americans, it fell in a class like that of the Mosquito, in which there was no comparable US type. But in 1949, this was not a problem for the Americans. It was totally a British airplane.

By 1950, however, the United States Air Force was in the market for an airplane to replace the aging Douglas B-26 Invader. With its sudden involvement in the Korean conflict, the Air Force was caught short with bomber types left over from World War II. United Nation's air superiority allowed curtailment of enemy supply lines during daylight hours, but it was at night that the North Koreans had little resistance to movement. The only bomber suited to the night interdiction role was the B-26, and night attack with these nimble aircraft was purely visual.

As a B-26 night intruder pilot in Korea, in the early stages of my Air Force career, I felt that the most important aspect to a newer airplane was for it to carry more bombs and be able to remain longer over the supply routes in the target area. We had heard of a jet bomber being contemplated as a replacement for the B-26, but we could see little use for a faster airplane. There was also promise of sensing equipment that was under development to be used for non-visual night interdiction operations, but to use this sophisticated equipment properly, there was obvious need that a more advanced aircraft should be acquired first. An even more urgent reason was to have a replacement for the dwindling inventory of B-26s which, at wartime attrition rates, was forecast to be depleted sometime in 1954.

Eventually, the concept of Air Force needs in terms of the next light bomber was focussed on night interdiction duties. This was in the summer of 1950, and a committee of Air Force officers was appointed to evaluate all available British, Canadian and United States aircraft that might be quickly adapted to the night interdiction role. To assure dispatch, the selection was to be made exclusively from existing designs, since creating a new type would add years to development time. With this as a prime prerequisite, few could foresee that problems lay ahead which might well take longer to solve than those associated with a new design. Aircraft of existing types to be considered were the Martin XB-51, of which two were flying, the North American B-45 Tornado, already in the Tactical Air Command inventory in substantial numbers, the North American AJ-1 Savage, a composite jet and reciprocating-engine bomber, designed to operate from large Navy carriers and already in fleet service. Foreign designs included the A. V. Roe Canada Ltd, CF-100 Canuck, a twin-jet all-weather interceptor, and the

In the early 1950s, the United States Air Force was badly in need of an aircraft to replace the Douglas B-26 Invader. This was the A-26 light bomber of World War II called upon for Korean War duty in which its numbers were being exhausted.

The Martin XB-51 was designed to meet certain military requirements, only to have them changed when the need for a night intruder materialized during the Korean War. This tri-jet came close to winning. (Martin)

The CF-100 built by A. V. Roe Canada Ltd was a very new fighter nearing production status. It was soon learned that it did not have the potential for being a jet bomber, yet those who flew it gained much in evaluating the all-weather and night flying qualities that could be built into such an aircraft.

English Electric Company's Canberra. This light bomber with its fighter-like maneuverability and speed, had possibilities!

After several gruelling months of evaluation by the board of Air Force officers, the final selection was to be made with flight demonstrations and comparisons at Andrews AFB outside Washington, D.C. in November 1950. The delayed arrival of the Canberra caused this date to be changed to 26 February 1951. This Atlantic crossing by the Canberra added even greater publicity to what was in store for the fly-off at Andrews AFB. Departing Aldergrove, Northern Ireland on 21 February, the crew of three RAF officers headed for Gander, Newfoundland. Since this was an RAF bomber, it was thought to be more appropriate that an RAF crew make the flight rather than the demonstration pilot already slated for the fly-off. The 1,785 nm flight was made in 4hrs 40min giving an average ground speed of 383kt. This was an unofficial record time for an Atlantic crossing in either direction, and was also the first direct, unrefuelled Atlantic crossing by a jet aircraft. A few days later on 24 February, the aircraft flew from Gander to Andrews AFB, the

North American B-45 Tornado had promise in the night intruder role, but its structure was based upon World War II design. Two engine nacelles housed its four jet engines. With four engines it had good partial engine performance. Pilots rated it easy to fly.

Another North American product in the evaluation was the AJ-1 Savage. This was already in fleet service but showed little growth potential for the Air Force requirement. Augmenting two R-2800 radial engines was one J33 turbojet buried in the aft fuselage area. (North American)

For several weeks, bad weather across the North Atlantic delayed the arrival of the new Canberra in the United States. Anxious spectators crowded around the sleek bomber as it came to Andrews AFB, near Washington, DC on 26 February 1951 where it was to be demonstrated. Its reputation for performance had preceded it and to many, it was a foregone conclusion that it would be selected by the USAF.

The 19ft chord of this broad Canberra wing is dramatically demonstrated in this view taken soon after airplane's arrival in the United States. The low stance to the ground simplified the plane's maintenance appreciably. (Martin)

point of the intended demonstration. While there was little doubt that the Canberra would out-perform its contemporaries, its flying demonstration was deemed necessary to silence those opposed to accepting a foreign aircraft into the inventory. Furthermore, a strong faction felt the Martin XB-51 was the best selection to make, but no firm decision could be made without a rigorous fly-off.

The flight routine for each aircraft at Andrews was to be a tight turn in each direction over the spectators, a slow speed and high speed pass and finally a short field landing. Elementary as it may sound, all of this was to be accomplished in a maximum of 10 minutes, a difficult task if the aircraft was not highly maneuverable.

Flying the British entry was former RAF Wing Commander Roland Beamont, test pilot for English Electric, and the Canberra test program. He was a master now in the Canberra, having flown the attention-getting show at Farnborough in an earlier version. His first reaction to the flight schedule was extreme disappointment, for it imposed an unreasonable limitation for demonstrating the full capability of the Canberra. Quick to see a solution to the problem, Beamont realized that nothing was said about use of any time left over from the allotted 10 minutes. The Canberra was easily capable of performing the routine in nearly half the prescribed time, while competing aircraft with heavier wing loadings such as the chief contender, the XB-51, *could not complete* the sequence in time.

To start the fly-off, the North American B-45 roared off the runway and into the distance, leaving twin thick smoke trails as it climbed almost imperceptibly into the clear morning sky. Eventually it reappeared, thundered past the spectators and homed in on the opposite horizon. Soon it returned and performed what some

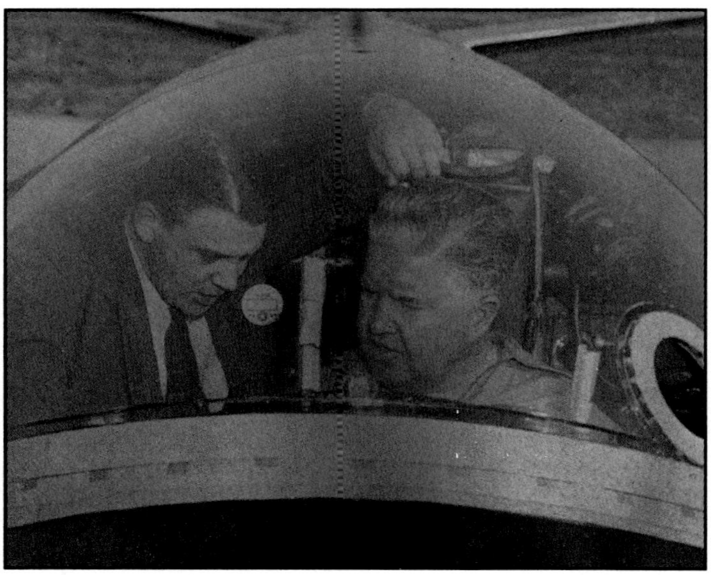

While the British Canberra was being considered for American acceptance, Martin's Chief of Flight Testing, O. E. 'Pat' Tibbs (right) is shown familiarizing himself with gadgetry on the English bomber with Wg Cdr Roland P. Beamont (left) English Electric chief test pilot. An RAF crew ferried the Canberra to the US but Beamont did the demonstration flying. (Martin)

An innovation used on both the British Canberra and the American B-57 were these speed-brake fingers. These protrude on both the top and bottom of the wing to help reduce speed. On the American B-57B and later models, these were supplemented with fuselage-side speed-boards.

Four landing flaps provided considerable surface area. These were two-position hydraulically operated and air-load balanced.

interpreted as the programmed 360deg turn. But its time was already up and observers showed signs of impatience while waiting for the Tornado to land so the demonstration could continue.

The AJ-1 Savage and Avro CF-100 were far more agile and accomplished the routine in respectable fashion. Next, the Canberra's prime contender, Martin's XB-51 circled overhead and was seemingly unnoticed by the crowd. It did not land at Andrews, but proceeded back to its starting point at the Martin plant at nearby Baltimore. Now the real showstopper headliner was on stage and

The Rolls-Royce Avons on this early model of the Canberra were started by a powder-charge cartridge positioned in the nose of the engine. Later Canberra models carried three cartridges, while the Americans when using the Sapphire retained the single shot starter charge but of a larger size.

To those who saw the Canberra perform, there was no doubt that this new bomber was superior to all other aircraft in its class. The Americans needed a new jet tactical bomber, one selected from existing designs. The best plane was one being built by a foreign power, fortunately an ally. There was no common ground for comparison, for the Canberra was a replacement for England's legendary Mosquito, another aircraft without an American counterpart. (Martin)

onlookers knew it. The star performer was the English Electric Canberra. Its low clearance to the ground and broad, yet stubby wing, were not pleasing lines compared to the more accepted swept wing designs, but in the air, its performance overshadowed any cosmetic shortcomings. After a short take-off roll, the bomber rotated at 80kts, and quickly assumed a 45deg climb at 150kts. After sufficient altitude was gained, it wheeled around, and still at full power, flashed past the crowd at nearly 500kts. With what sounded like a dual flame-out, power was chopped to decelerate and the plane was brought around into the first 360deg turn at a tight four Gs within the bounds of the airfield. Then a reverse in direction was made for the second turn, having advanced full power again, holding 2.5 Gs at 200kts. After completing this second turn in about half the width of the airfield, wings were levelled and the Canberra shot up to an altitude where the gear and flaps were extended and it spiralled down to the scheduled slow fly-by at 110kts. This ended the program with three and a half minutes remaining before the landing had to be made. Taking advantage of this time, Beamont snapped up the gear and flaps, applied full power and zoomed back to a thousand feet over the heads of the reviewing officers. Gear and flaps were dropped again and a tight 360deg landing pattern was executed which put the bomber on the runway with one-minute to spare.

This spectacular showing ended with some embarrassment, but the situation seemed ignored by many in attendance. At the point of touch-down, white sand that had recently been put down on ice went unnoticed by the pilot. With heavy braking pressure for a short run landing, the left wheel momentarily locked and blew the tire. There was no damage, and some spectators regarded this as a further demonstration of the forgiving nature of the airplane.

This demonstration of aircraft at Andrews only helped to confirm a decision that had been all but finalized a few weeks earlier. The Canberra had been chosen as the most suited for the night interdiction role. The just completed aerial demonstration had been staged ostensibly to support this decision – a decision obscured by politics and one which was not overwhelmingly supported by those involved in the final selection.

3

THE EVALUATION

The committee of officers selected to evaluate the airplane candidates spent six months on the project. Their objective was defined: first, match an existing airplane that can best destroy tactical targets at night and under bad weather conditions, one with the capability to destroy stationary and mobile targets with conventional and atomic weapons of sufficient size, and lastly; be capable of photographic and electronic reconnaissance. A more defined yardstick was developed with performance figures for use in measuring an aircraft's potential for the night intruder and tactical bomber/reconnaissance role.

Yardstick for night intruder (1950 concept)

Take-off ground roll:	3,000-5,000ft
Rate of climb:	5,000-8,000ft/min
Cruise, loaded:	350-450kts
Service ceiling:	30,000-40,000ft
Max speed:	450-550kts
Target area time:	6min bomb, 6min evasion
Bomb load:	6-10 500lb GPs
Guns:	6-10 .50cal or equivalent
Rocket capability:	May be carried
Radius of action:	800nm

Unfortunately, each aircraft evaluated had been designed for a different mission objective. Consequently, for this test, each had to be evaluated in relation to the mission for which it had been designed. Its adaptability to the light bomber role was then to be considered. The evaluation committee was more than casually interested in Canada's new Avro CF-100 Canuck, the prototype of which had just flown a few months earlier in January 1950. This was a two-seat all-weather and night twin-jet fighter, and had many of the qualities desired for the new tactical bomber. Because of its small size, however, they concluded that the Canuck could not carry a sufficient bomb load without a major redesign of its structure. Further, the maximum range that could be designed into the airplane was far short of that required for the bomber.

Just a year before the introduction of the CF-100, North American came forward in May 1949 with the AJ-1 Savage as a Navy carrier-based strategic attack plane. At the time it showed great growth possibilities, promise which failed to be fully realized. The plane had two R-2800 radial engines and a J33 turbojet located under the tail. The Savage had many of the qualities as a light bomber that the evaluation team was looking for. It could operate well from short unprepared fields and had satisfactory range and loiter capabilities for the intended mission profile. Its bomb load capability

The CF-100 was most unusual in design with its thick center section joining the two 6,000lb thrust Avro Orenda turbojet engines. With addition of tail warning devices to detect attacks at night or in weather, the evaluation committee graded the CF-100 able to survive combat encounters well beyond its anticipated operational life span. One measuring yardstick was ability to defend against attack by F-86 type fighters as seen in the background. (Avro)

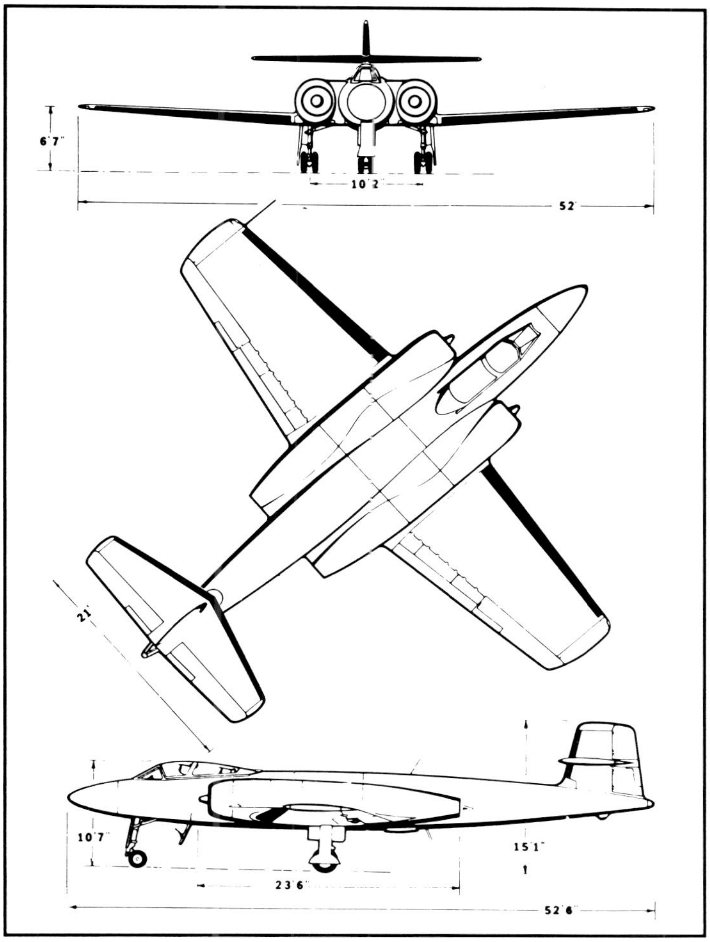

Avro CF-100 Canuck.

was greater than required, yet the Savage retained its maneuverability to perform night attack missions. It proved to be a very stable bombing platform, a feature the committee was highly conscious of. The cockpit gave good all-round visibility to the side-by-side crew, except in rain, when droplets would just hang on the windshield. Interior layout was excellent for night operation and all-weather flying which matched well with the plane's stability. Unfortunately, the AJ-1 was one of those transitional aircraft that did not have the speed and performance being demonstrated by the new family of jet aircraft presently under development. This, coupled with the complicated hydraulic system, made the evaluation committee feel that the craft would be too vulnerable to aircraft and ground fire. Another objection was that this navy bomber did not carry forward firing or defensive armament. North American engineers gave assurance that this could be remedied but the team members felt that such a major redesign would add penalties to the plane's performance, already marginal.

Consideration of the English Electric Canberra as a candidate for the night intruder mission was based on an earlier examination of the airplane in England by Air Force officials. In August and September 1950, this group led by Brig Gen Albert Boyd, of the Air Materiel Command, had gone to England and been enthusiastic over the new bomber. Few disputed that England was more advanced in jet technology than any other nation. Furthermore, unlike others, the British were proud to share their finding with closely allied nations. This superiority was based upon England's production of jet engines with greater thrust and durability than those of U.S. jets. After enumerating the good and bad features of the Canberra, Gen Boyd and his committee concluded that by USAF standards, the British aircraft seemed best suited for the following three roles, in order of preference: all-weather fighter, tactical reconnaissance vehicle, and medium-high (25,000-40,000ft altitude) short range bomber. It is hard to envision this bomber as an all-weather fighter, but when examining the qualities needed for this type of mission, the Canberra possessed many of the key prerequisites. Among these were high altitude, stable platform with speed and range making it superior to any current or contemplated USAF all-weather fighter for combatting B-50 and B-36 type targets. Ironically, the recommendations pointed out that it was doubtful that the Canberra could be employed above 35,000ft because of acceleration limitations, a supposition which was to prove erroneous.

In the short range tactical reconnaissance role, the Canberra was able to operate at altitudes beyond the reach of a great majority

North American offered an alternative to the AJ-1 Savage, the A2J-1 featuring turbo-props and other refinements. Its development could incorporate the Air Force requirements, but the prospect was shelved in lieu of more advanced designs. (North American)

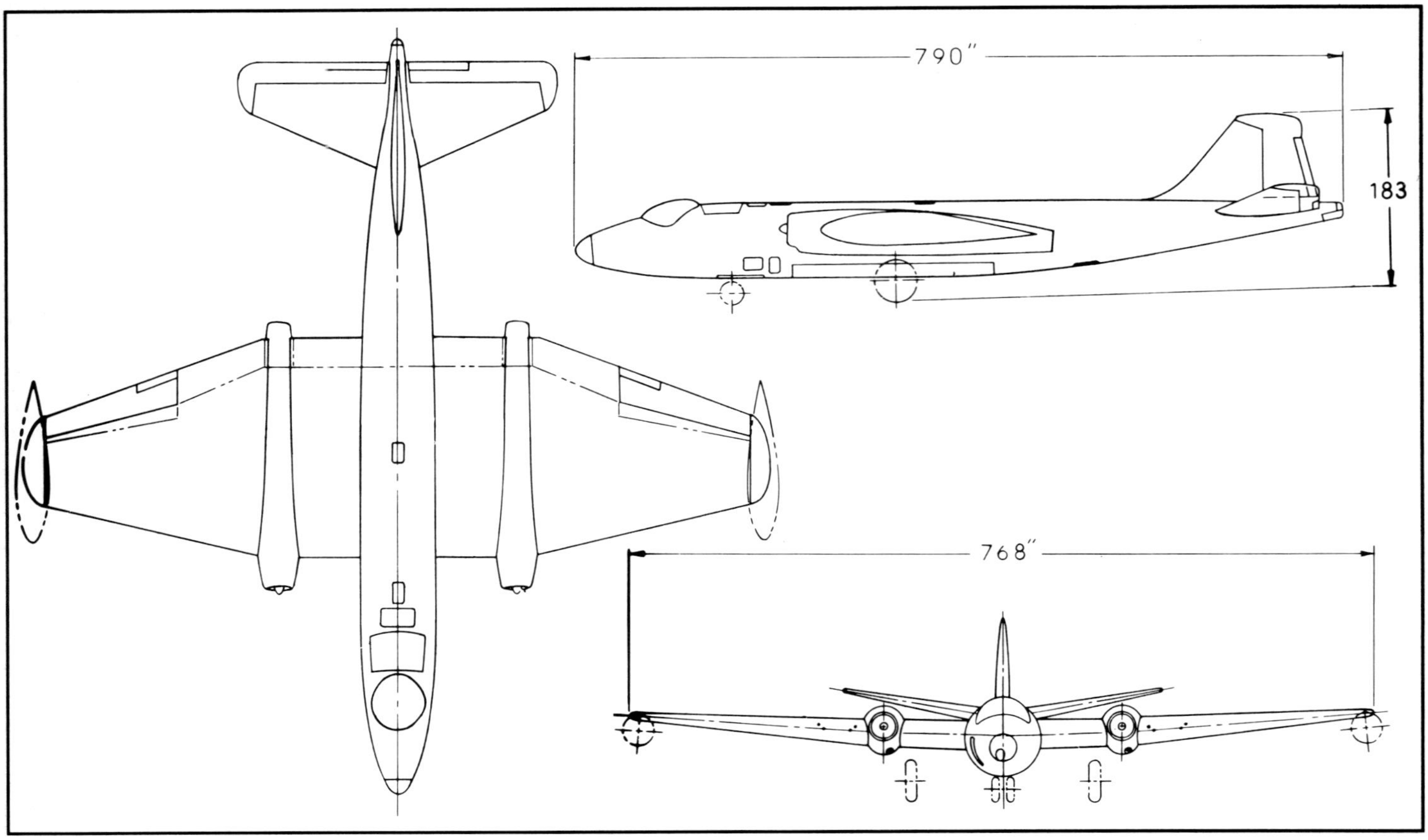

English Electric B2 Canberra.

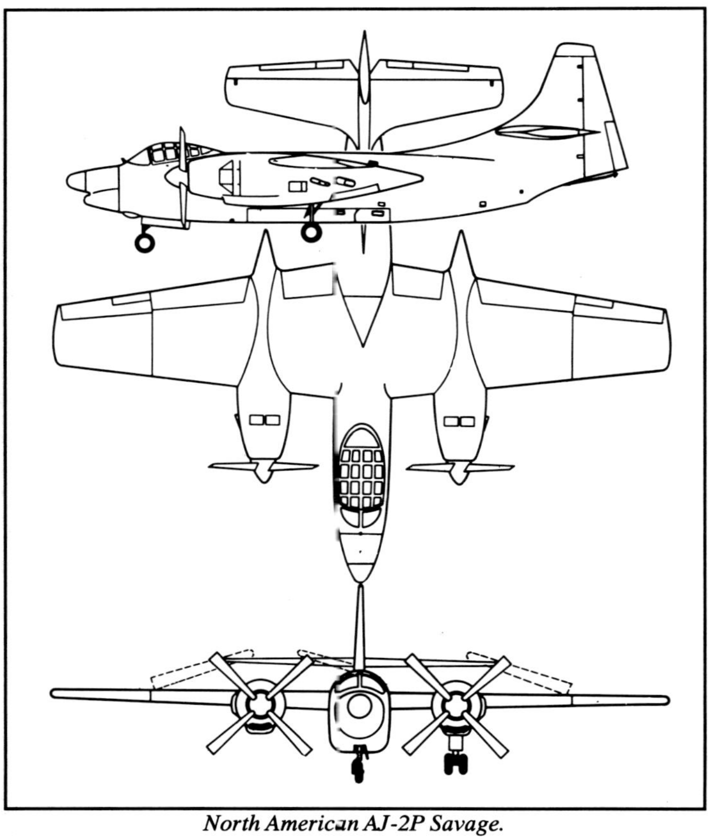

North American AJ-2P Savage.

of then operational interceptors. Finally, the success of the Canberra as a medium to high altitude, short range bomber seemed to hinge primarily on the capability of the navigation-bombing system to be incorporated into the airplane. The lower the altitude of the penetration and bombing run, the Canberra would be more vulnerable to MiG-15 type interceptors. In addition, speed would also have to be reduced on the bomb run to minimize bomb bay turbulence, further penalizing the Canberra.

On a purely economic point, the reviewing group felt that a Canberra selection would have considerable merit by virtue of its incorporation into the MDAP (Mutual Defense Assistance Pact) program. In that case however, it should be integrated into the USAF only after the successful conclusion of a 'rigorous evaluation' of at least one aircraft and accelerated service testing of 10 service test models. At this point in the selection process, if the Canberra was, indeed, a potential American bomber, it would be purchased directly from British production lines for the U.S. portion of MDAP participation. Given this set of ground rules, the Canberra must have really impressed the visiting selection board, especially when one considers how hungry U.S. manufacturers were in the five years immediately following World War II.

One very pertinent observation from the committee noted that: 'It is mandatory that the USAF accept the general airframe, performance, and load carry capability as is in order to retain the advantages accrued as a result of the design philosophy used in the Canberra aircraft.' In theory this is a desired concept, but in any new airplane, changes are inevitable and this group noted that if finally procured in production quantity, the airplane should have

Early English Electric Canberras like this B Mk 2 were powered by Rolls-Royce Avon 101 engines of 6,500lb thrust as compared to the J65 Sapphire with 7,200lb of thrust built under license from Armstrong Siddeley which powered the Martin B-57. The Canberra design was void of defensive armament, relying on its speed and maneuverability to escape interception. Martin's B-57 was developed from this variant.

no less than 25 specified changes incorporated into it to make it suitable for USAF needs.

Now the qualities of the Canberra were to be evaluated for its merits in the night intruder light-bomber role based on the Boyd report being reviewed by the committee. Taking into consideration the many good features the report noted, committee members felt, however, that the Canberra's greatest deficiency would be in target tracking – either air-to-ground or air-to-air, because of its light wing loading causing an unstable platform in turbulent air. Minute corrections would also be hard to make while following the bombardier's command. This may be valid in theory, but in actuality it did not hold true for this airplane. From my own flying experience in the B-57, it was a very 'rough rider' in turbulent air due to its non-flexing wing. More than once my lap belt could not be tightened enough to keep my head off the canopy, but turbulence did not set up an oscillation effect and it remained stable for easy target alignment. Bomb run corrections at altitude were often 'stiff', but practice and flight instrument interpretation compensated for this. Obviously, the committee did not have this actual experience with the Canberra from which to correct these assumptions obtained from the earlier report.

The Canberra's crew facilities were rated marginal. The pilot had sufficient cockpit working room but overhead clearance was shallow. The navigator, located behind and below the pilot, scarcely

Since the Canberra was one of the very early jets of this relatively large size, its low stance with the ground seemed most unusual since propeller clearance was no longer a problem. This gave great ease to ground maintenance of the engines and other accessories. Cockpit entrance was one long step up to floor level through the nose hatch under the canopy. (BAC)

With the acceptance of the Canberra design for American usage, one of the first recommended changes was to add a drag chute to the tail cone. It became evident that this would be unnecessary for an aircraft able to stop after a 1,300ft roll at near empty weight, half that of the chief contender, the XB-51. Another unfulfilled prediction from armchair engineers was that the fin and rudder would be enlarged if accepted for American production. (Martin)

Martin's entry, the XB-51 is shown just after take-off, with gear and flaps coming up. XB-51 was unusual in that its wing could pivot on the fuselage to increase its angle of incidence when the wing flaps were extended. This was necessary because of the bicycle type landing gear that prevented the aircraft itself from making a pitch change angle for take-off. (Martin)

had room in which to work. His having to move forward to operate the bombsight, met with disfavor, and it was also questionable if the Norden bombsight, still standard in the Air Force, had sufficient space to be installed and operated.

Should the Canberra be selected for the USAF and purchased directly from British production lines, changes would have to remain minimal by both necessity and Air Force directives. However, the essential change recommendations included: (1) adding forward armament, (2) Shoran bombing system, APW-11 bombing-aid radar guidance system, and a suitable gun/rocket/bombsight, (3) standard Air Force instruments and lighting, (4) Air Force oxygen system, (5) engine and airframe anti-icing, (6) correct stick force gradient from a push force to a pull force in the landing configuration at speeds below 120kts, (more on this later) and (7) suitable cockpit ventilation for low level summer operations.

Despite the Canberra's obvious strong points, the apparent favorite throughout the evaluation with others in the Air Force was Martin's XB-51 of which the first of two was flown in October 1949. The original mission of this aircraft was low level attack on surface military targets in support of ground forces. Although this experimental bomber was 'caught in the middle' when the primary mission for the Air Force's new light bomber was changed, the XB-51 seemed a natural for the new role, even if it became difficult to match it to the requirements set down for the night intruder aircraft.

The XB-51 was a burly machine. It had two J47 jet engines of 5,200lb thrust, mounted on the lower forward side of the fuselage

The tri-jet XB-51 incorporated a drag chute for emergency short field landings, a feature so well liked, that it was recommended to be included in the American built Canberra. The proposal proved unnecessary. Note the slight aileron and massive landing flaps on the XB-51. Spoilers were primarily used for lateral control. (Martin)

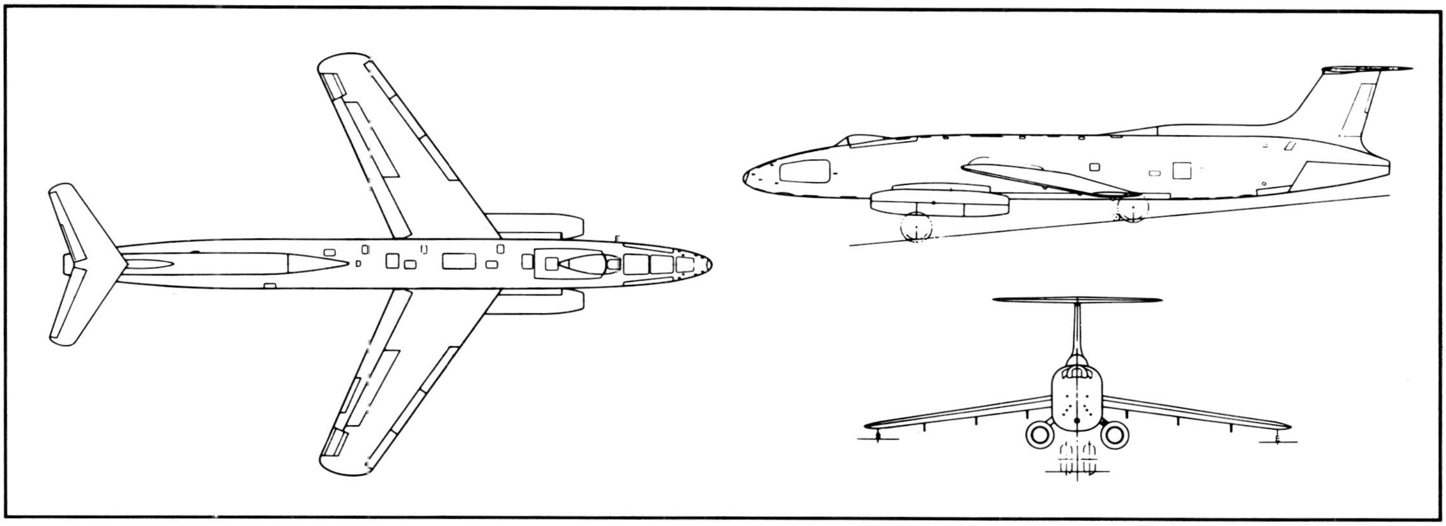

Martin XB-51.

and a third J47 in the tail. Having a tandem (bicycle) type landing gear, there was not the normal pitch change for take-off and landing. A variable angle incidence to the wing operated when the wing flaps were extended. Another unique feature was the first time use of a rotatable bomb bay door on which the bombs were mounted. When open, its load was the same as external stores, but without speed restrictions, up to its maximum allowable for the airplane of .93 Mach or 540kts IAS, whichever was reached first.

This Martin entry was a highly maneuverable airplane for its size. At low levels it had a very satisfactory turning radius in the speed range of 280-310 IAS, giving it excellent versatility for night bombing operations. However, its low limit load factor of 3.67Gs severely limited its capability during tactical operations and was graded unsatisfactory in this area of evaluation. Furthermore, despite a forecast of improved jet engines under development, there was little hope that the range and endurance of the XB-51 would improve sufficiently to meet the night intruder loiter time requirements. The XB-51 was not set aside quickly, however, for it incorporated many features to its advantage for a tactical bomber. These included single point refuelling, rebombing in nine minutes, rearming in 15, and low footprint weight, allowing it to operate from hastily-prepared airfields. In combat configuration it could operate if need be from 6,000ft runways at sea level, provided there were clear approaches. Its all weather characteristics appeared to be satisfactory. Despite being a dense airplane (crowded with actuating systems and other components) it was rated as being easy to maintain. In addition to its novel rotatable bomb bay door, it had pneumatic assist bomb release to assure positive release at high speeds. The small dimensions of the bomb bay however, severely restricted the load and ordinance varieties that could be carried.

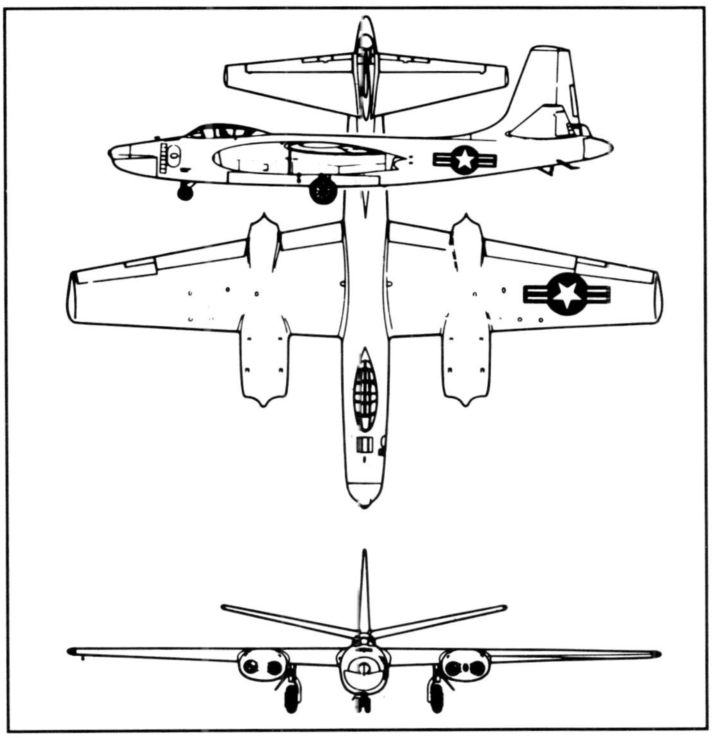

North American B-45 Tornado.

From a defensive standpoint the level flight speed of .89 Mach below 30,000ft for the XB-51 made interception by then present-day fighters in the F-86 class extremely difficult. Its armor protected engines, remotely mounted from the basic structure, fire extinguisher system and fuel tank purging was all highly regarded as reducing vulnerability to aircraft and ground fire.

When the evaluation board came to the North American B-45, they did not make a formal investigation, as time was growing short and several on the team had flying experience in it as a TAC bomber. The Tornado had been around for several years following its first flight in March 1947, and more data was available on it than any of the other entrants. In general, those who flew it felt that the handling characteristics of the B-45 were satisfactory in the high speed range which was limited to .77 Mach, but the three Gs at that speed imposed a limitation on maneuverability. In the speed range of 250-300mph IAS, it was a nimble airplane for its size and considered capable of performing the night low level work satisfactorily. It had a surplus of loiter time available in which to survey the target area for interdiction work at the lower altitudes. While existing models were not equipped for gunnery or rocketry, a number of simulated attacks had been made in the B-45, and it was considered satisfactory for that type of work.

The B-45 had excellent crew accommodation. Its main drawback was the restriction of forward vision for the co-pilot by the front pilot's ejection seat, due to the tandem arrangement. Crew of four was carried, including a gunner in the tail compartment armed with two .50cal machine guns.

The Tornado had a great number of good features that matched the night intruder 'yardstick.' Unfortunately, its design was based upon World War II concepts and it was already the oldest of the types being considered. It was really too big and too heavy to be well suited to this role. However, as a stop gap measure for the Korean War, the committee did make the recommendation that existing B-45s in the TAC inventory be transferred to the night interdiction mission for the calendar year 1951. This was never done, however. Perhaps the reason was that at this time, heavy combat losses were being experienced in Korea with the few reconnaissance versions of the B-45 that were deployed there.

Armed with these evaluations of the five airplanes (some of which are shown in Appendix 1), the committee submitted their findings to a Senior Board of Officers on 15 December 1950. Instead of one, they picked two bombers as having potential in the night intruder role; the Martin XB-51 and the English Electric Canberra. Their proposed plan was to purchase immediately Canberras from England as an interim airplane to equip two of the contemplated four light bombardment groups of the newly restructured 95 Wing Air Force. This would allow time to produce the Martin B-51 to equip the remaining two groups, then to re-equip the first two Canberra groups with B-51s as well.

This recommendation was short lived, for the Senior Board did not agree fully with the way the evaluation yardstick was applied to available aircraft and believed that the AMC committee was prejudiced in favor of the B-51. They fully recognized that the B-51 was nearly 100kts faster than the Canberra, but the British bomber was far superior in flight endurance. The Canberra with its two 6,500lb thrust Rolls-Royce Avon engines allowed it to loiter for approximately two and a half hours over a target 780nm from its base. The B-51 even with future engines could loiter but one hour over a target 350nm from base. Concern was also expressed that the wing tip wheels on the B-51 would prove troublesome at hastily prepared forward bases, while the Canberra was best suited to this environment over all aircraft considered. Of importance also was the fact that the Canberra was the only light jet bomber contemplated for use in the then-current NATO (North Atlantic Treaty Organization) program of which the U.S. was a key member.

The board could not unconditionally recommend procurement of the Canberra above all others until it obtained more information from the British as to their ability to supply, over their own needs, 300 aircraft which would make up the four U.S. bomb groups. Should the Canberra become available at an acceptable rate, the board was very much aware that there would be a number of modifications necessary before it could fill the role of a night intruder bomber. Also recognized were the inherent problems of stocking and maintaining components of English measurement standards for replacement items, and using British type accessories. This would create a logistic problem, for all support would have to be established as a system separate from normal USAF supply channels. This uncertainty of purchase, the many modifications needed, the failure generally of the airplane to meet USAF standards, indicating the probability of operational problems, further confused and prolonged the issue before the selection board. Thus, for a more critical and penetrating evaluation they requested the immediate loan of one completely equipped airplane to be brought to the United States. The RAF agreed and one Canberra was set aside for the flight to the U.S. This airplane was WD932 described in a previous chapter and now handed over to the USAF on 5 March 1951.

While awaiting these final tests, the question of airplane availability, should the Canberra be selected, become a major issue with the board. Lt Gen K. B. Wolfe, Air Force deputy chief for materiel, and a group of AMC officers visited England to obtain first hand production data. They soon learned that the British were willing to supply the Canberra to the United States, but deliveries could not be accomplished at a suitable rate for the USAF and still meet RAF commitments. A cost was quoted at $1,474,000 for each of the 300 Canberras that were to be contracted for.

An alternative for meeting production needs and the factor that saved the Canberra for the Air Force, was British willingness to grant manufacturing rights of the Canberra to the United States. However, with this prospect of major U.S. production involvement, Gen Wolfe and his group warned that the Canberra did not meet Air Force aircraft standards and that much of the proposed modifications necessary for U.S. production would entail serious problems. Since the Canberra was built to the British system, an entire redesign to U.S. standards would be necessary. It soon became apparent, however, that this was the only solution should the Canberra emerge as the winning choice. Anticipating U.S. production, the board recommended that a contractor be selected to manufacture the Canberra in the U.S.

The Glenn L. Martin Company was approached with the proposal to build the Canberra for the USAF should the XB-51 lose in the selection as a tactical bomber. They willingly agreed, for this assured them a much needed contract in either case, though it was evident they preferred to build an airplane of their own design. It was during this period of indecision that the designation 'B-57' was assigned to the Americanized version of the Canberra bomber. Within the Martin Company, this would be the Martin Model 272.

When the long-awaited decision making day in February 1951 finally came as the first Canberra arrived at Andrews AFB, members of the Senior Officers Board, Headquarters USAF, TAC and ADC were on hand to appraise the new bomber. On 26 February, the ground and aerial comparison between the vying aircraft types took place, and shortly thereafter the final decision was made. 'The Canberra,' the official board report noted, 'comes closest to filling the night intruder profile because of its excellent characteristics of endurance, range, maneuverability, and the visibility provided from the nose section.' Thus, on 23 March 1951, a letter was sent to the Martin Company requesting that 250 'B-57As' be manufactured to fill the night intruder bomber role for the USAF. The U.S. had purchased and was preparing to manufacture its first foreign military designed aircraft in 35 years, dating back to the S.E.5 and De Havilland DH-4 of World War I vintage.

4

CANBERRA - B-57 COMPARISON

My first close look at this new airplane came at Langley AFB, Virginia in October 1954, when O.E. 'Pat' Tibbs, director of flight and Chief Test Pilot for The Glenn L. Martin Co, flight demonstrated the B-57B to our unit. This aerial exercise was far more spectacular than the one given at Andrews AFB in February 1951, the one which clinched acceptance by the USAF of the Canberra design. Tibbs was not confined to a set or timed demonstration as was the case at Andrews, and the broad range of speed and maneuverability of the airplane was unbelievable unless seen. That afternoon, following the morning demonstration, Tibbs held an introductory and background briefing about the development and flight characteristics of the B-57. This evolved around his flight testing of the English Electric Canberra that was initially on loan to the Martin Company for that purpose. Reading from prepared notes, Tibbs' captivating and dry-humored story began to unfold:

> Reaching approximately 30,000ft with the Canberra, I increased the speed until it reached .83 Mach. At this point, because of severe roughness, I decided to discontinue further investigation of speed as I was somewhat dubious about the British method of determining limit buffeting values. It seems to me that probably it was based on the point *just before the airplane completely disintegrated.'*

Pat's briefing went on after our uneasy laughter, covering our concern, had subsided:

> 'Now, the B-57 aircraft you see on the line and that you saw flying this morning resembles the Canberra I just described when I evaluated it in 1951. But believe me, gentlemen, there is no resemblance in the performance today of the two aircraft as far as the high speed characteristics are concerned.'

The alleged difference was not by accident. Originally, when the Air Force purchased the design rights of the Canberra, they had felt that little or no time for further aerodynamic 'clean-up' of the Martin-built product would be needed to match the *stated* performance of the British Canberra. However, with the advent of jet engines, the time had come where power out-put was exceeding the structural capabilities of airframes. In the case of the B-57 and British Canberras, if left at full throttle, especially at the lower altitudes, the airplane would reach a speed where it would disintegrate in a matter of a few short minutes. This point was being met at a low Mach number because of air flow buffeting before the aircraft reached the stipulated speed limitations, and the consequences did not augur well for the new Canberra. The British tried to cure this roughness at high speeds by increasing the strength of the fuselage skin. Tibbs jokingly described that approach as 'merely armor plating it, mainly just so they could live with the roughness.' But there was underlying concern in his voice. The cause of the problem had to be detected before the right solution could be made.

In tight formation with another B-57, Tibbs watched the lead aircraft as they approached and exceeded .81 Mach at 40,000ft to see where the trouble might be. The test aircraft had been provided

One of the finest demonstrations of the Martin B-57B to crews who were to fly them took place at Langley AFB, Virginia, on 11 October 1954. The author is in this group about to have a closer look at the Canberra following its aerial demonstration. At this point, the Canberra was still a curiosity with members of the 345th Bomb Group, for it was several months before the new bomber was assigned for operational use.

Seasoned Martin test pilot O. E. 'Pat' Tibbs demonstrated the full operating limits of the Canberra is such a way to make any skeptic take notice. Everything about Tibbs with test flying was professional, and if Martin built it, 'Pat' flew it first. Demonstrations were not 'air shows', but merely to show the reliability and capability of the airplane. They were still spectacular!

The short field take-off capability for the Canberra was impressive for any bomber. Flight demonstrations were often without wing tip tanks and a light fuel load, making the take-off distance less than 2,000ft, depending on wind and temperature. In this picture taken just after breaking ground, the wheels are already beginning to retract.

with a series of strings or tufts secured along the skin to visually indicate areas of turbulent air. By the gyrations of the tail, the problem area was obvious. 'I couldn't help but wonder how much of *that* an airplane would withstand before tail feathers would begin to part,' Tibbs continued. 'I loosened up my formation as a matter of self-preservation while we came back on our airspeed!'

A temporary fix was tried by covering and wrapping all the openings in the tail and aft fuselage section with tape. In the case of the B-57 design, it had an adjustable stabilizer used for trim. To allow for this movement, a large open slot at the attachment point to the fuselage was necessary, and this was the area of most turbulence. Air was going in through the fuselage members, reversing, and so forth, and was probably causing the trouble. A second flight test was made which revealed a marked improvement. All B-57s were then equipped with snug-fitting covers for these gaps, yet free movement of the stabilizer was allowed. Aft fuselage bulkheads were also sealed to further resist free passage of air in and around the inner structure. This was acclaimed as the greatest contribution by the United States to aerodynamic design improvement for both the British as well as the American built Canberras.

The obvious question was often asked at the conclusion of most of 'Pat' Tibbs' briefings: 'What differences did you note in comparing the flying characteristics of the British Canberra with that of the initial B-57As which were hardly distinguishable from the British counterpart?'

Prepared with an answer, Tibbs often started like this:

> 'Due to the Americanization, many think we have lost much performance over the British airplane. They usually indicate that they are sure we have installed a lounge, a bar, a ready room and so forth aboard, which would completely ruin the airplane. Well, let me say right now, nothing could be further from the truth. I can detect no difference between the two aircraft in take-off, climb out, cruise, etc – no changes. At first ours was just as rough, if not more so, as theirs until cleanup. Control forces are slightly lower on the B-57s than on the Canberra. This very desirable condition probably stems from a better aileron sealing job obtained in converting from British to American standards, plus the fact that the three control surfaces – elevator, aileron, and rudder – have slightly lower spring rates in the tabs.
>
> 'The greatest advantage with our airplane was the Martin rotatable bomb bay door. The Canberra has a speed limitation of 350kts IAS for opening its bomb bay, even though the doors slide up inside the fuselage. Ours could be opened at its maximum indicated airspeed of 500kts IAS with no adverse effect. In general, the performance of the B-57 is equal to or better than the English Canberra in every category that we have tested up to this time (1954).'

Tibb's opinions were not always shared by his English Electric counterparts, however. Tibb's overstatement in his briefing on performance was understandably for morale and sales purposes, but comparisons as stated were accepted as being exaggerated. According to Roland Beamont, former chief of flight test for English Electric, when reiterating this story to him about the differences in the two aircraft had this to say:

> 'As you know, I flew flight envelope and performance tests on the '57A and B models and found level performance identical and no improvement on buffeting from the gap-seal modifications which we considered to be cosmetic. All Canberras at Warton were tested to .84m after the prototype Canberra structure was cleared to .86m. Above that, a strong nose-down pitch occurred which could not be held until recovery below 15,000ft and at lower Mach. The buffet above .83 was heavy, but caused no problems. Control pressures were the same with both British and American models.'

Flight demonstrations were no easy task as evidence by 'Pat' Tibbs' sweat-soaked flying suit after landing at Langley. The B-57 did not have control boost and was a heavy airplane to maneuver at low altitude. The ground crew seemed more interested in removing nose hatch to reload starter cartridges for the next start, than to steady the makeshift ladder for Tibbs.

Following the morning flight demonstration and a well earned shower, 'Pat' Tibbs discussed in great detail with future Canberra flight crews the handling qualities of the B-57. Of interest to most listeners was his description of differences between the British-made Canberra and the Martin B-57. This was the briefing room of the 500th Bomb Squadron at Langley in October 1954 at which the author was in attendance.

For some inexplicable reason, the B-57 suffered a nose-up trim change at a lower Mach number than its British counterpart. This was only cured by fitting a small spoiler along the full span of the upper surface of the horizontal stabilizer. According to G.M. Hobday, English Electric's representative at Martin at the time:

'Without exception, the clean-up of the Martin built product including the stiffening of the rear fuselage, the sealing of the aft fuselage bulk heads and the stabilizer gaps evolved from British work and was in fact passed on through me to the Martin Company. There was in agreement that any improvements in design by either party should be passed on to the other.'

5

Production and Setbacks

In order to reach this point in the development of the B-57B that was just demonstrated at Langley, many transactions took place over the preceding three and a half years following the Air Force acceptance of the Canberra design. On 9 March 1951, the Air Materiel Command (AMC) implemented the Canberra procurement and production program by authorizing the purchase of an initial quantity of 250 Canberras with FY 1951 funds. A license agreement between the two companies was consummated on 3 April 1951, covering the manufacturing and sale and use of the Canberra and spare parts solely in the U.S. to and for the Government. Royalties were not to exceed 5% of the selling price of the airplanes at fair market values.

To assist Martin in production, five subcontractors were to produce about 60% of the airplane by airframe weight. Kaiser Metals was to build the two wing panels and nacelles, as well as the rotating bomb bay doors. Hudson Motors furnished the aft fuselage section and the tail components, while Cleveland Pneumatic Tool Co supplied the main and nose landing gear. Other sub-contracted items included the canopy and nose cap, tip tanks, fuel cells, forgings and castings, ejection seats, hydraulics and electrical equipment. The engine was an American adaptation of the Armstrong Siddeley Co Ltd Sapphire engine having 7,200lb st. This was a more powerful engine than the 6,500lb st Rolls Royce Avon that powered existing Canberras, but the extra thrust would be needed for the anticipated added weight of the B-57. Wright Aeronautical Corp paid $499,800 for manufacturing rights to build the J65 Sapphire engine. Buick Motors Division of General Motors Corporation, sub-contracted and initially produced the engine for the B-57s.

The first Canberra drawings arrived from England on 1 June 1951 and work began immediately on the conversion from British

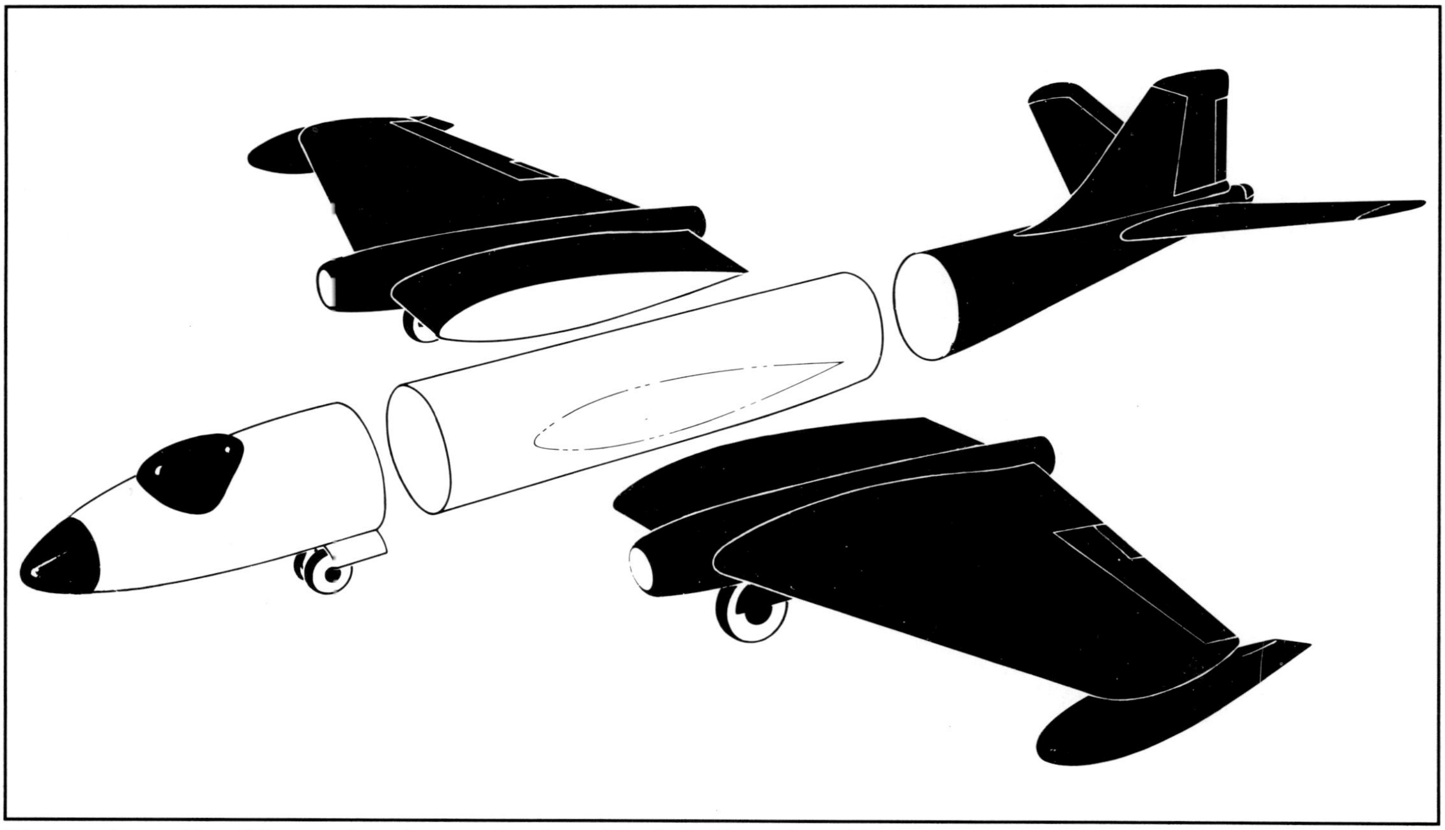

Diagram gives an idea of the extensive subcontracting planned for the B-57 as indicated by the shaded components. Nearly 60% of the plane by weight was to be built outside the Martin plant. Kaiser Metals fell short of its wing production and their work had to be accomplished by Martin. Hudson Motors built the aft fuselage and tail section. There were five subcontractors in all for these major components.

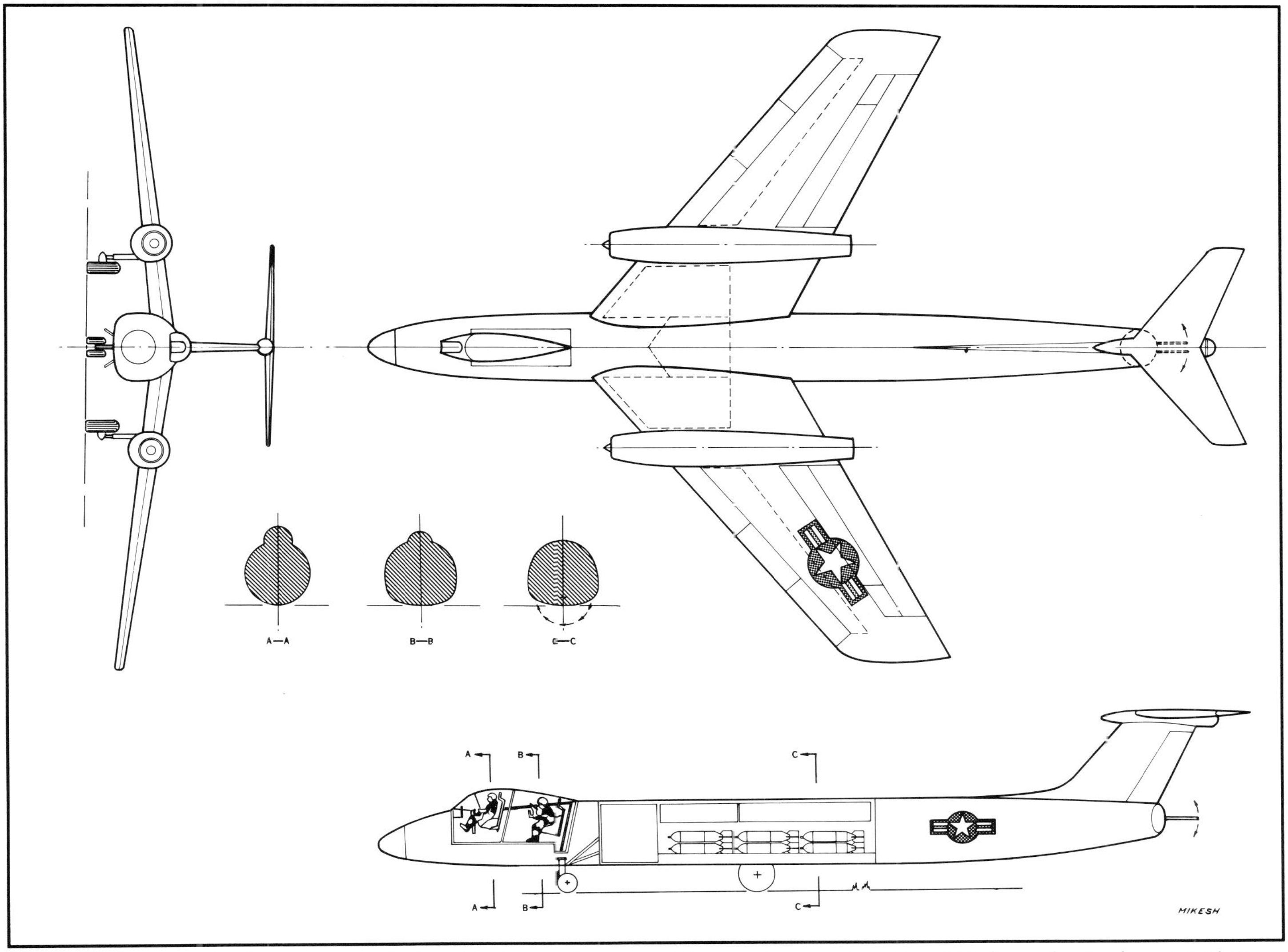

Proposed B-57B Super-Canberra was never built mainly because it was virtually a new design and was thought to have taken too long to put into production. Borrowing many features from Martin's own XB-51 as well as the Canberra, it promised much better speed and performance.

measurement standards to those of American standards and related equipment. This entailed a tremendous redrawing job, for the Americans used different gauge sheet metal standards, rivets, bolts and screws, and the best compromise had to be determined in each case. Manufacturing tolerances in production methods were also a major consideration. Instead of being required to hand-tailor details in assembly to ±.020, ±.005, the tolerances were increased to ±1/32in. This in no way affected the quality since outside mould lines were still maintained in final assembly. Aside from production method changes, there were inherent design changes that had to be made. To what limits this should be carried was a sensitive subject that persisted for many months, for prerequisites made upon the Air Material Command when accepting the Canberra for the USAF was that it would be built 'as is' with minor exceptions. Major changes would increase costs and extend production delivery schedules and the urgency created by the Korean War could not tolerate this. On the other hand, this 'no significant change' policy generated considerable doubt and apprehension among certain Air Force agencies, particularly the newly implemented Air Research and Development Command. Their mission conflicted with the selection criteria of the Canberra 'as is,' and although the research and development agency was responsible for the technical excellence of Air Force aircraft, its hands were tied in the case of the B-57. They determined that 'the Canberra fell far short of meeting existing USAF requirements and will require major redesigning to be brought up to what could be considered satisfactory." Thirty-five deficiencies were listed, but to stay within the guidelines of the purchase of the airplane, only six of these were corrected. Many felt and rightfully so – the aircraft's usefulness 'as is' within the USAF would be extremely limited and could only be considered an interim measure until a suitable design could replace it.

The Martin Company, at the request of the Wright Air Development Center, recognizing this frustration, which they too shared, offered to Headquarters USAF an entirely new design for a tactical bomber and night reconnaissance airplane. The new proposal was called the B-57B Super Canberra. In general, the new design was a combination of the best features of the XB-51 and the B-57A. Flight experience gained from the XB-51 provided a sound footing upon which Martin was able to develop this proposed high performance tactical bomber. Basically, the Super Canberra was a swept wing

Martin employees watch expectantly as English Electric test pilot Roland Beamont delivers the Canberra to Martin on March 6, 1951, where production will soon begin at this Baltimore plant. Martin had hoped to build their XB-51 for the Air Force but the Canberra contract was better than no contract at all. British markings stayed on this aircraft for the several months of test flying while it was on 'bail' to Martin. (Martin)

At a cost of $1,018,388, Martin purchased two Canberra B2s from English Electric, and the money was reimbursed to Martin by the USAF. WD932 had crashed by the time this transaction was completed and WD940 was the only one of the two to have its serial number 117352 applied to its fin. (Martin)

design, retaining low wing loading and embedded engine nacelles like that of the Canberra. Improved Sapphire engines with afterburners would provide a total of 21,700lb of thrust at take-off and a maximum speed of 630kts at sea level. Fuselage diameter was the same as the Canberra and its T-tail was a carryover from the XB-51. The crew of two sat in tandem under a single teardrop canopy, while photo reconnaissance and ECM missions could be accomplished with bomb bay package conversions.

Martin's engineering report of 14 September 1951 claimed that production airplanes of the Super Canberra could be available for service during the calendar year 1954. However, a new and unproven design is what the aircraft selection board wanted to avoid, and by November 1951, the Super Canberra proposal was rejected. In the meantime, flight studies continued with the Canberra that was on bail to the United States from the RAF. Following its arrival at Andrews AFB in February 1951 for the final evaluation and

The first step in re-marking the U.S. Government owned Canberra was to remove the RAF roundel and apply U.S. national insignia. This pattern is not a paint mask, but merely served as a guide around which to apply masking tape. (Martin)

bomber selection, this Canberra B2, WD932, was flown to Langley AFB, VA, for members of Headquarters TAC to have a look at their future night intruder bomber. From there it went to Wright-Patterson to be inspected by AMC before being flown to the Martin factory at Baltimore.

Pat Tibbs made the delivery flight to Martin's Middle River, Maryland, facility where the B-57 would be built. Before leaving the plant for the pick-up of the airplane, Chet Pearson, president of the company, made these parting comments to his Chief of Flight Test. 'Pat, if you like the airplane at all, do a slow-roll or something across the field when you return. We all will be waiting. The morale is pretty low here for us, having designed our own plane, yet have to build a foreign design.'

Tibbs liked the airplane, and by the time he reached the Martin airport, he was confident of its feel and maneuverability. He gave the folks at Martin what they wanted – and then some. Ten feet off the deck, Pat brought the Canberra across the field at 400kts indicated. At the far end with the roar of sound that just reached the spectator's ears, he pulled it up steeply and went into an aileron roll. On its back during the roll, there was a sudden silence as both Avons flamed out – and for the next few minutes the Canberra was a high speed glider. Fortunately the light fuel load remaining gave Tibbs the time needed to get the first re-light at 800ft followed by the second, after which he gingerly brought the plane around to a conventional and graceful landing. Cool thinking, with a good working knowledge of the equipment, and having the situation always in hand is the prime ingredient of a competent test pilot like Tibbs. Few besides Tibbs were the wiser about what had just taken place and the people at Martin were a bit happier at building the English Canberra.

By June of that year, the need for a second Canberra was recognized, and approval was given by AMC to Martin for the purchase of the loaned aircraft at hand, and one yet to be delivered, at a reimbursable cost of $1,018,388 for the two. The arrival of the second Canberra B2, WD940, from its North Atlantic crossing on 31 August 1951, generated more interest than the earlier crossing of WD932. This second airplane established an official time record from Aldergrove to Gander of 4hr 18min which averaged 417kt.

Beamont, of English Electric, who had performed the flight demonstration at Andrews AFB six months earlier in WD932, was in command for this crossing. With Beamont were navigator D. A. Watson, and radio operator R. H. T. Rylands. On 4 September 1951, this second Canberra joined the first at Martin for the Combined Test Project Agreement between the USAF and the RAF which called for the exchange of aircraft development information.

WD940 was the 12th production B2, and the 21st Canberra to fly. After nearly a year at Martin, WD940 had its insignia changed from that of the RAF to US markings, but retained its light gray upper surfaces and black lower surface camouflage. Across its tail was painted its 1951 assigned Air Force serial, consisting of the numbers '117352,' but it, like WD932 were directed to retain the identity of 'Canberra' and not classed as 'B-57'. Tests made with this Canberra were limited to four 48,000lb max gross weight take-offs to simulate the initial gross weight to be designed into the B-57A. These heavy-weight take-offs could not be performed on the first aircraft because of its wheel and brake limitations. When WD940 ended its last flight and was placed in storage, its total flying time was 33hr, of which Martin pilots flew it for only three of those hours. Eventually the landing gears were interchanged between the two airplanes so the earlier and more fully instrumented WD932 could handle the remainder of the flight test program.

Martin pilots flew the first Canberra, for 41hr out of its total 86hr as of 15 October 1951. It retained its RAF insignia throughout the test program and its Air Force serial number (51-17387) was never applied to its fin. The reason for the number not being applied was that it had not been assigned until after the plane had been destroyed and was then belatedly paid for.

Tragedy struck the Canberra project when this airplane, WD932 crashed while on an evaluation flight just before Christmas 1951. On 21 December, during a tight turn analysis at 10,000ft, the left wing failed just outboard of the engine nacelle. The airplane crashed near Centreville, Maryland, on the Delmarva Peninsula, 25 miles south-east of the Martin factory. Both crew members ejected, but the engineer-observer's chute failed to open, and he was killed. Air Force test pilot Maj Harry N. Lister received minor injuries.

Investigation revealed that structural failure had occurred when the pilot was pulling 4.8Gs at 420kts at the aft C.G. limit of the aircraft as the test called for. Later investigation of the accident suggested through the process of elimination that improper fuel control consumption from the No's 1 and 2 tanks caused the C.G. to move aft at an alarming rate and was in fact far aft of the specified aft C.G. limit when the aircraft broke up. (The aircraft was stressed to an ultimate factor of 7-1/2G and reached this loading with some reserve during structural tests in England.)

Naturally there was concern over this accident and structural failure back at Warton. Canberra test pilot Roland Beamont duplicated the flight to confirm the original test conditions while the investigation was underway. He reported that 'I re-proved this case at Warton a few weeks later in another production B2, WD958, to 5.2G at 450kts, before we knew the results of the analysis. Subsequent calculations showed that with incorrect fuel management on that fatal flight, the C.G. could have moved well aft of the aft limit

Rolled out from Martin's paint shop, RAF WD940 is now USAF Canberra 117352. B-57 designation was not used with British built aircraft. Probe on left wing tip is yaw indicator for flight evaluation only. Open hatch leads into upper electronic bay, forward of the main fuel cell. The number 2 painted on the nose was Martin-applied to distinguish between the two Canberras while both were at the Martin factory.

The British Canberra cockpit was to be adequately functional, but found to be quite cluttered by post war design standards. Lever on control column was brake handle.

This is the mock-up cockpit for the American designed interior for the B-57B bomber version. Simplicity and well grouped systems controls was of prime concern in the design.

and when pulling the G-force tests, the aircraft could have pitched up to 6-1/2 to 7G.'

While the investigation into the cause of this accident was underway, it provided the leverage needed to make design changes in the B-57A design. There was little doubt, though unconfirmed, that escape from the aft navigator's station was critical as was suspected through this fatal accident. Neither crew member carried out the correct ejection procedure, which in this particular early aircraft was to manually blow the canopy and navigators hatch before seat ejection. Both the pilot and navigator ejected *through* the canopy and hatch respectively. Some opinions were that the engineer observer would have lived if he had not been knocked unconscious, since he died from drowning and his injuries were in fact quite minor.

In January 1952, the Commander of the Wright Air Development Center, Maj Gen F. R. Dent took a positive stand on the shortcomings of the Canberra design for use in the USAF. In a letter to his command headquarters, he noted no less than 31 design deficiencies, pointing out that if all could not be eliminated in the existing design there would have to be a major redesign of the airplane. To reinforce his stand, General Dent enclosed a document which the British Ministry of Supply sent to the English Electric Company. This contained a long list of flaws which had to be corrected before the Royal Air Force would accept the airplane. The general noted that the British deficiencies '. . . very closely correspond to those outlined by WADC in presentations prior to the Headquarters, USAF letter directing production of the B-57A.' Furthermore, the general continued, 'It appears inconsistent that we should ac-

This striking picture is of the re-marked English Electric Canberra to illustrate the new USAF bomber to be produced as the B-57 night intruder. The morning sun reflects on the Chesapeake Bay, and the nearly completed Bay Bridge in July 1952. The bridge dates the picture for this vital link to the Delmarva Peninsula which is now a double bridge system.

Sporting USAF markings, the former RAF Canberra WD940, was the holder of the record for an east-west transatlantic crossing – 4hr 19min, from Aldergrove, Northern Ireland, to Gander, Newfoundland, on 31 August 1951. Eventually this airplane was used by subcontractors for patterns.

The setting sun did not halt production of the B-57, spurred on by the war in Korea. Two work shifts were employed by Martin, until the first four bomb groups were outfitted. RB-57A in foreground is being towed to the paint shop for its coating of anti-searchlight black paint, while B-57A, ship No 5 (52-1422) at left, already has its markings and was left in natural finish. (Martin)

cept these same deficiencies in the production model of this airplane for use by the USAF.'

These are but a few examples of the problems that faced the Canberra in its conversion to the B-57. According to G.M. Hobday, previously mentioned in connection with the early design of the Canberra, these so called thirty-one design deficiencies were largely of a political nature aimed at curtailing the use and production of the aircraft for the USAF. The deficiencies in fact were very largely the differences between the 'MIL' SPEC and AP970, which were the respective technical requirements for the design of aircraft for the two air forces.

Hobday was in a good position for making these observations. He had been the resident advisor engineer to The Glenn L. Martin Company in Baltimore from July 1951 until August 1954 and the sole representative in the U.S. for English Electric in the redesign of the Canberra as the B-57.

Enlarging further on the political tangle that the Canberra/B-57 was caught-up with, Hobday offered this further explanation:

> 'I can quote to you at least two examples where redesign took place at great expense, and of course a waste of time which did nothing to improve the B-57 over the Canberra in any way whatsoever.
>
> 'The first of these was the re-design of the main wing spar to the fuselage attachment from a fatigue point of view. The British design stood the test of time and was a simple two-bolt affair. The USAF model, on the other hand complicated this attachment joint immensely and it became a four bolt affair.
>
> 'Secondly, there was the matter of flying control problems. The British aircraft had a simple push pull system which was designed to meet two requirements, that of stick force exerted by the pilot, and also one of stiffness. The American requirement was based purely and simply as a load produced by the pilot with no stiffness requirement whatsoever. Total re-design of the system using cables was undertaken for the B-57 to 'MIL'SPEC requirements. Now let me add this – I have never known any pilot, including many American exchange pilots, who have flown the British Canberra ever break the controls due to shear physical effort!'

The fate of this airplane was continually bombarded with political opposition from the original date of the decision to select the airplane through to its last production order. There was in both the Martin and British view a sizable degree of opposition both at the technical level at Wright Field and in many other quarters. The full potential of the aircraft production run once set and tooled up for fifteen hundred aircraft was never reached, largely as a result of this opposition.

6

DEVELOPING THE NIGHT INTRUDER

With further study of the problem facing the success of the B-57A as a successful night intruder bomber, AMC recognized that there did exist an incompatibility in producing the airplane 'as is.' Although production was already under way, Martin was asked to re-engineer the problem for consideration and approval. What resulted from this study was the B-57B, incorporating modifications to the basic Canberra design. This was a second use of the designation since the earlier concept of the 'Super Canberra' no longer existed.

On my first visit to the Martin plant to have a look at the mock-up of the newly designed nose for the B-57B, I felt immediately that this version of the Canberra was genuinely an American airplane. I remained skeptical of the wide chord, low aspect ratio wing which was not comparable to any other airplane of any other nation. The wing imbedded engine was not duplicated in any other American jet and we who were to fly the B-57 had some reservations about this. Engine failures and fires could cause disastrous wing structure damage, but very few aircraft were lost through causes that might be attributed to this design feature.

Tandem Canopy

The new cockpit arrangement, though weighing 387lb more than the original, was a remarkable improvement for the pilot. (Little did I suspect then that I would spend over 2,000 flying hours under this type of canopy!) Visibility needed for the ground attack phase of the bomber mission was nearly unrestricted in all directions. Of equal importance, it moved the navigator from the deep compartment behind the pilot with only one small window on the port side, to a position where he could see out of the airplane.

This new canopy – formally requested of Martin in May 1952 – was essential to correct a deficiency that was overlooked when the Canberra design was first selected. The original double curved layers of glass would flex with changes in temperature and pressure, especially during the ground attack phase of the mission. It therefore became impossible to place a gunsight behind this canopy and have proper harmonization with the guns for acceptable accuracy. This sight had to be placed behind a flat glass panel.

The second British Canberra to be received, WD940, was used to work out these new bomber-version features. It was modified

Former Canberra WD940 was modified by Martin with this new tandem canopy for engineering study and was possibly used for flight evaluation. This led to the B-57B. The clear forward nose section was left unchanged for this evaluation. This elongated canopy adds even slimmer lines to the Canberra. (Martin)

Mock-up of Martin's redesigned nose section is evaluated closely by Wright Air Development personnel. This model became known as the B-57B, its cockpit reflecting American styling throughout. This redesign was necessary in order to provide a windshield with flat glass behind which to place a gunsight without having distortion due to temperature and pressure changes. (Martin)

Bombs could be pre-loaded onto the bomb bay door and rolled to an awaiting B-57 for rapid turn-arounds when carrying complicated loads. Four hand-held hoists lift door in place for attaching to its pilot points in bomb bay on which the door rotates 180 degrees open and closed. Wheel dollies were then removed. (D. Anderton)

with this new cockpit arrangement which also required the moving of flight controls and pilot position from the off-set left of center, to the centerline location. If WD940 was ever flown in this configuration, it is not known. If it had, it did not make an impression on 'Pat' Tibbs who would have flown it after being modified. WD940 was then made available to sub-contractors for inspection of the components they were to make, and later was reported as being moved to the museum yard at Sampson AFB, Geneva, NY where it was later scrapped on 11 September 1956.

There is evidence indicating that this tandem canopy arrangement for the Canberra may not have been an entirely "Americanization" of the design. According to G.M. Hobday, English Electric advisory engineer to Martin for the conversion process, he had this statement to offer in this regard: 'The Night Intruder development which became the B-57B resulted from an in depth design and study by English Electric including wind tunnel work, which was cancelled by the British Government. I was in fact chief technician at the Acton/London Office of English Electric where the new cockpit was designed, immediately prior to going to Baltimore in 1951. The concept therefore was not exactly thought of in the USA.' Since the British concept was not fully developed to production, or now in physical evidence, the similarities, or dis-similarities of the two must remain a mater of conjecture.

Speed Brakes

The new 'B' model was to have another improvement feature – speed brakes at the waist position of the fuselage. It was learned from the earlier design, that although unusual, the finger-like spoilers, or wing dive brakes (as they were technically referred to) on the top and bottom of the outer wing panels did not provide sufficient drag for speed control. At the high operating altitudes, idle power settings retained a higher rpm with resultant high thrust in the thin atmosphere. Since cruising speed at high altitudes was equal to maximum allowable, reducing power and extending the wing

This dramatic picture was taken at near bomb release point in the LABS maneuver. When just past the vertical position the atomic special weapon was released and continued on its own to around 9,000ft before starting down toward the target. At release, the B-57 would continue in an Immelmann-turn and begin its escape before the bomb would detonate. B-57s were the most accurate and stable aircraft for this critical bomb delivery method.

CUT CANOPY FOR
EMERGENCY RESCUE

OPPOSITE: The first B-57B shows off its new canopy and its improved cockpit vision becomes obvious. Windshield framing restriction to forward visibility was hardly noticeable. Overhead lines in rear canopy was antenna for ADF. When positioning the special ladder for crew access, it was aligned with the two 'U' shaped marks on the side of the fuselage. Flat black on leading edge surfaces was rubberized paint which resisted wind erosion. (USAF)

dive brakes, had only minimal effect. The let down from altitude became tedious and time consuming so as not to exceed the airspeed red-line. These wing dive brakes were retained and worked in conjunction with the controls for the fuselage speed brakes. In the ground attack role, the speed brakes would be very useful for controlling acceleration in diving passes.

Wing Guns

Modifications to the wing of the B-57B were necessary in order to house four .50 cal. forward firing machine guns in each wing. These were placed in a common gun bay in each wing outboard of the engine. The newest weapon at the time the 'B' model was being developed was the fully automatic, air-cooled, recoil-operated Browning M3 .50cal. machine gun, a variant of the M2 that was so widely used during World War II. The M3 had an increase rate of fire to 1,200 rounds per minute and was made to be more adaptable to lower temperatures because of the higher and colder atmospheres of which jet aircraft were to operate. In addition, each gun was electrically heated and pneumatically charged upon demand by the pilot while in flight.

The author removes the expended starter cartridge at the end of the delivery flight of 551 to the National Air and Space Museum. These cartridges were for a one time start and dependent upon good electrical contact points despite the dirt from the burning material that this system emitted.

Each of the eight machine guns had 300 round of ammunition for a total of 2,400 rounds for the airplane. Ammunition was fed to the gun in a disintegrating metallic link belt. Muzzle velocity was 2,900 feet per second. When considering an average single burst of 2-seconds, this would send 320 rounds into the target.

After the 83rd B-57B, 52-1575, the remaining bomber versions (which was the majority), were equipped with a total of four wing-mounted 20-mm M39 cannon in place of the eight .50 cal. machine guns. These were gas-operated, belt-fed, electrically-fired, percussion-charged weapons. This newest of the airborne weapon systems not only equipped B-57s, but F-86Hs, F-100s and F-101s as well. Machine gun equipped B-57Bs were not converted to the cannon weapons.

On cannon equipped B-57Bs, the guns were fixed to fire downward at 3 degrees 36 minutes from the flight path and converge at a point 3,250 feet in front of the airplane. The intent was designed for convoy strafing in a lesser dive angle than when guns were fixed directly forward. Each gun could fire 290 rounds of ammunition, totaling 1,160 rounds for the airplane. These guns had a rate of fire in excess of 1,500 rounds per minute, putting 200 rounds of this larger ammunition on a target in one 2-second burst. Usually the rounds were armor-piercing-incendiary (API). Each gun weighed 179lbs and was 72.4in long.

To ready the guns for firing, the armorer had to charge each gun separately three times prior to flight. This was normally done in the arming area near the takeoff end of the runway since the guns could not be charged in flight. One disconcerting aspect of the 20mm cannon system was the slight time delay after pressing the gun switch. The gun purge doors opened hydraulically when the gun switch was depressed before allowing the guns to fire. Initial reaction was that of having a gun malfunction.

Bomb Bay Door

A feature often overlooked in marking the success of the B-57 tactical bomber is its rotating bomb bay door. This one piece, 17ft long door was mounted on two pivot points. It could open in four seconds and close in six, making a 180deg turn. The bombs were attached directly to this door, and in the open position the bombs would be in an externally mounted position. This kept the bomb bay cavity sealed or closed, therefore causing no buffeting or pitch change generally associated with conventional bomb bays, and open doors that often affected bomb dropping accuracy. This eliminated any door opening speed restrictions, which was at 350kts for the British Canberra. Developed first on Martin's XB-51 and carried over to the B-57, this door innovation was invented and patented by two Martin Armament Engineers, Albert T. Woollens and Werner Buchal. In addition to its performance advantages; the removable doors could be pre-loaded at a remote site, then towed on its own detachable wheels to the airplane for a quick mission turn around. No one recalls this loading procedure being used when actually being prepared for combat missions in Vietnam. Presumably it was

The engine start of a B-57 was immediately obvious by 10 seconds of black smoke that belched from the starter exhaust of each engine. The right engine was started first to build-up hydraulic pressure so that the canopy could be closed to protect the crew from the left engine starter smoke. Clean burning starter cartridges were later developed. (USAF)

easier to manhandle loading the bombs with the door in place than interchanging the door.

Weapons Delivery

Features of this door contributed to the B-57 becoming the most accurate special weapon delivery aircraft through the 1960s for the unique bombing system called LABS (Low-Altitude Bombing System). With no speed restriction for opening the bomb bay, this allowed a high speed entry for the maneuver close to the ground (about 50ft) at 425kts. With the door already open, a 3.5G pull-up was made at the target, and the special weapon was automatically released at about the 110deg position of the loop. The bomb continued up to about 9,000ft while the airplane was going over the top at about 5,000ft. This 360deg vertical-maneuver could be completed or an Immelmann-turn made at the top, depending on which escape direction was desired away from the target. Gaining speed again on the downward side of the maneuver at full power, the B-57 would be a considerable distance from the target before bomb detonation. B-57s were more accurate than any other aircraft for this type of delivery.

A Shoran bombing system operated by the navigator/bombardier was included, as well as a APW-11 Bombing-Aid Radar Guidance System for the pilot. For detection of other aircraft, an APS-54 Radar Warning System was included. This provided an audible warning to the pilot when an airborne interception was in a position to offer a potential threat to the airplane. Eight hard-points were added to the underside of the wings for bombs and rockets to be attached.

Other Features

Some of the features that were recommended early in the Canberra program that did not materialize were, wing surface and engine inlet anti-icing, anti-skid wheel brakes, drag chute for landing on short runways, power boost controls, center-line mounted wing tip tanks (similar in design to those on T-33s), and an AM-1 target avoidance warning radar which was an indicator that told the pilot when minimum pull-out altitude was reached. Only on very rare occasions did I have need for any of these features, and their absence did not detract from the effectiveness of the airplane.

Of interest also on all J65 engine equipped B-57s was the unique starting system that used an electrically ignited single-shot starter cartridge. When fired, it burned for 10sec, directing its force against a starter turbine. This turbine drove the engine through a clutch linkage system and brought the engine up to starting speed. During the process, early cartridges emitted a dense black smoke, indicating to the uninitiated that the airplane was on fire. More than one of us got hosed down by the unsuspecting fireguard during early days of operating the B-57. The purpose for this method of starting allowed airplanes to be dispersed in a combat situation, doing away with heavy ground powered starting units, difficult to transport and

For the production of the B-57, Martin moved back into their former government-owned World War II B-26 Marauder assembly plant No 2 across the road from their main facility. RB-57As are shown in various stages of production. Navy personnel served as purchaser and handled administrative matters for the US Government and USAF at the Martin plant. (Martin)

Martin's Chief of Flight Test, 'Pat' Tibbs, climbs aboard the first Martin built Canberra for test flight at the Baltimore Middle River facility. George Rodney at left went along as Engineering Recorder for the flight. 'First Flights' for Tibbs at Martin were nothing new. Flights of this nature went back to pre-World War II with Marylands, Baltimores, and every Martin type that followed. (O. Tibbs)

The smile on 'Pat' Tibbs' face is proof enough that the first flight of the Martin B-57A went well. One recommendation was to improve the air conditioning system for prolonged low altitude flight, as evidenced by Tibbs' sweat soaked flying suit. The large plastic canopy trapped radiant heat and there was only limited, unfelt, outside ventilation. (O. Tibbs)

maintain in out of the way places. What engineers overlooked however, was that the B-57 was designed with a liquid oxygen system. This system had to be recharged every 24hr or before each flight, and required a factory-like facility close at hand in which to generate the liquid oxygen.

The J65 Sapphire engine proved to be a very reliable engine for the B-57. In retrospect, however, some feel that this may not have been the best choice over that of the Rolls Royce Avon. The Avon proved its undoubted superiority and had greater power. Its method of starting by using smokeless cartridges from the beginning and the triple breach configuration was superior to the American system. The selection of the Sapphire may have been a matter of availability and on more favorable terms than the Avon.

The first B-57A breaks ground in the early morning sun as the wheels are almost fully retracted. Tibbs reported after this flight that the flying qualities of the Martin built B-57A were no different than that of the English Electric Canberra. First flight of the B-57A was 20 July 1953. (Martin)

21438
21432
25
21
19
18

U.S. AIR FORCE
21442
U.S. AIR FORCE
21438
U.S. AIR FORCE
21436

Ship No 9, 426, the first RB-57A taxies out for acceptance flight in 1954. Last of the eight B-57As, 425, in silver finish behind, was instrumented for Shoran Bombing tests that were later conducted at Eglin AFB, Florida. None of the eight B-57As were considered combat aircraft. (Martin)

Production

When the design was approved for the B-57B night intruder configuration, the entire B-57 production schedule had to be revised. Since production was already underway for the B-57A, only eight would be completed, having slight alterations incorporated. No longer was this initial configuration expected to perform the bomber functions. Components that were already on the production line for others were converted to the reconnaissance craft as RB-57As. These were to be limited to 67 in all, and now, reconnaissance production by necessity, was ahead of the bomber version. In order to adjust for the added engineering and retooling costs for the bomber revision, the total quantity of aircraft for the fixed price contract for FY1952, was reduced from 250 to 177 which set the figure in August 1952 at 102 B-57Bs, including spare parts, a mobile training unit and 103 special weapons doors.

First Flight and Deliveries

The day every manufacturer awaits, finally arrived for the people at Martin on 20 July 1953. Just 28 months after the awarding of the Canberra contract, the first B-57A took to the air. 'Pat' Tibbs was at the controls and he later reported that the Martin airplane handled no differently than his earlier flights in the British Canberra.

***PREVIOUS:** A line-up of RB-57As on the ramp at the Martin factory in 1954 receive final adjustment before being turned over to the Air Force. Most of these pictured went to the 363rd TRG at Shaw AFB, SC. In 1970, the author flew ship No 18 in foreground, (52-1435) on its last flight to the salvage depot at Davis Monthan AFB, Arizona. Ending with 15 years of flying, the Canberra could not have performed better on that final flight! (Martin)*

Before a gathering of top Air Force officials and some of the nation's leading industrialists, the first B-57 was officially turned over to the USAF on 20 August 1953. In attendance was USAF Chief of Staff General Nathan F. Twining who accepted the airplane after which Major Roy Seccomb made the first official USAF flight. This event was unusual in that it took place at evening twilight, in order that visitors might better visualize the mission of the B-57.

Getting the night intruder into production did not end here for this first airplane, as 52-1418 was not a combat machine due to recognized deficiencies in its design. Already there were three models of the B-57 following it down the production lines. Owing to the changes in production schedules it became a major problem to conclude a test program of 'debugging' before B-57s became operational. General Boyd, Commanding General of the Wright Air Development Center, let it be known to AMC that the B-57 situation was beginning to resemble the B-47, F-94C and F-89 programs, wherein these aircraft had been plagued by rashes of groundings, retrofittings, and openings of new modification centers. Generally speaking, Boyd continued, these situations could be expected when, '. . . an aircraft is accelerated to quantity production without adequate lead time for testing or correction of deficiencies brought out by development testing.'

As the schedule then stood, test agencies would receive RB-57As in November and December 1953, with deliveries to tactical units slated to start in December. Following this, B-57Bs were scheduled for test delivery in February and March 1954, with the start of deliveries to operational squadrons beginning in March. As a result of this time compression, experience gained from the test programs

Externally, the Martin built RB-57A showed no structural design changes to that of the British Canberra. The Martin-built Canberra contained their in-house devised rotating bomb bay door that eliminated buffeting when opened at high speeds over that of the clam-shell door type. Photo-flash bombs are shown attached to the door. (Martin)

could not be incorporated into handbooks initially delivered to the using agencies. There was no 'XB-57' for these evaluations since the British version was already considered a proven design. The problem was somewhat resolved by its own accord in February 1954, when existing production schedules broke down because of Kaiser's inability to meet wing delivery schedules. To resolve the production problem, Martin assumed the unfulfilled contract and built the wing panels in their own facility. This provided some measure of relief to the research command testing agencies because

Externally, the Martin built RB-57A showed no structural design changes to that of the British Canberra. Common to both are the dive brakes, later called speed brakes, that are square channels and emerged vertically from the upper and lower surfaces of the wing. Those on the right wing can be seen extended through the U.S. (Martin)

arrival at peak production rate of 20 per month was postponed from January 1954 to one year later, and delivery of the first B-57B was changed from February to July 1954. Contributing to production set backs was the slow delivery of the J65-BW-1 engines from Buick. Due to the tedious conversion task to American standards, they initially failed to meet Air Force specifications. By 1954, many B-57s sat on Martin's ramp awaiting engines with which to fly. Production of the engine finally reverted back to Wright Aeronautical Corporation as the J65-W-1 (later became W-5).

Before the first contract for 177 Canberras was completed, other production contracts followed. These changed repeatedly in quantities, price, and additional equipment. The first contract for FY1953 stabilized at 138 B-57Bs and an equal number of special weapon doors, plus spare parts and ground handling equipment. By June 1955 however, this contract was again amended to include 38 B-57C dual control aircraft and 120 more B-57Bs. This latter figure was reduced in October 1955 to 100, and the remaining 20 airframes were to become the B-57D. A third production contract covered the RB-57D series with FY1953 funds. The fourth and last, new production contract covered 68 B-57Es with FY1955 funds and brought the final figure of Martin built Canberras to 403 in six production line varieties. All were completed by February 1957 – over a time span of just three and a half years between the first and

For nearly two decades, the Martin B-57 Canberra was the only jet tactical bomber in the USAF. There was no mistaking the Canberra for any other airplane. Its wide chord wing compared to its mere 64ft span was its most distinguishing feature. With the Martin modified nose departing from the British design, its tandem seating under one canopy made it distinctively American in design. Only the wing imbedded engines were a feature not practiced in U.S. jet design. (USAF)

Designed to meet the night mission requirements of the USAF, all RB-57As and B-57B/Cs were delivered in overall anti-searchlight semi-gloss black. All leading edges were painted with a rubberized flat black designed to prevent wind erosion of the paint. All other markings were insignia red.

This is the once very popular color poster print that illustrated the new USAF bomber. These posters were widely distributed by Martin as public relations for their new product.

the last roll-out. In the interim, Gen Boyd's warning about accelerated production before 'debugging' became a reality. There were the predicted periodic groundings after fatal accidents, one grounding lasting nearly four months while awaiting corrective fixes.

Named 'Canberra'

Throughout the life of the B-57, there was a reluctance by those associated with it to call it the Canberra – its true and rightful name. Perhaps the reason stems in differentiating between the British and American product, and Martin's frequent reference to the B-57 in their manuals and news releases as the 'night intruder', has led some to believe this was its actual name. Paragraph 17 of the Letter of Agreement between English Electric and Martin, clarifies this point:

> 'Martin shall name all aircraft manufactured by it under this Agreement "Canberra" in accordance with the usual practice of Martin with respect to other aircraft of its manufacture and shall use its best efforts to procure the agreement of the Government of the United States of America that the same name shall be used by the Government of the United States of America.'

A check with the Air Force as to the official USAF name they carry for the B-57 reveals the record to show 'Canberra' for all models. This stemmed from the British tradition of naming many of their aircraft after major cities. In this case, the Canberra was named after the capital city of the commonwealth of Australia.

The service life for the B-57 Canberra as a combat bomber has passed. In retrospect, the question may still remain; did the Americanization of the Canberra go into production with the least number of changes – as directed – so as not to change the inherent design of the airplane? Or, did the other faction win out to the point that the Martin built Canberra became a pure American airplane, far removed from the British design except for general basic lines? I like to believe that it was a near perfect compromise of the best features of the two. Unfortunately, the promised sensing equipment for the night intruder role did not become available until the final phase of the Vietnam War when it was finally introduced into the B-57G series. Consequently, for the night intruder mission during the first 16 years of its operational life with outdated equipment, the B-57 was little more than a faster B-26. It proved, however, to be an effective replacement airplane for the Invader as attested by the fact that it remained in combat for eight years and had a total military service life of thirty-years.

7

The B-57 Enters Service

The first of the Canberras from the production line were the eight B-57As. For the most part, these became the test air frames for not only evaluating stability and control, airframe structures, but systems such as Shoran bombing, navigation and radio systems to name a few. RB-57As were the next aircraft to follow, and units scheduled to receive the Canberras were anxious for them.

These all-black reconnaissance aircraft were dispersed initially to a number of organizations. The first tactical unit to receive the Canberra was the 363rd Tactical Reconnaissance Wing at Shaw AFB, South Carolina, receiving RB-57As as early as March 1954. Others wasted no time being ferried across the Atlantic to two or more reconnaissance units in Germany; mainly the 1st TRS of the 10th TRW at Spangdahlem AB and the 30th TRS of the 66th TRG at Simbach AB. Although it is not clear, it is possible that some of these airplanes augmented other reconnaissance units in this wing and group as well as other units in different parts of Europe.

A slight variation to the reconnaissance model sent to Ramstein, Germany, were RB-57A-1s. This designation was applied after the fact to ten modified RB-57As that had specially selected J65 engines and given the name "Heart Throb." These engines were trimmed for producing added thrust. According to MSgt Clyde Scarboro, stationed at Ramstein at the time, these engines carried the RB-57A-1s appreciably higher than others, a need apparently

The first unit to receive Canberras was the 363rd Tactical Reconnaissance Group at Shaw AFB, Sumter, SC. Their RB-57As were readily recognizable by their red and white checkerboard tails. This was the only reconnaissance Canberra unit of the Air Force stationed in the United States. (USAF)

These B-57Bs of the 345th Bomb Group are on the break for landing at their home base, Langley AFB, Virginia. Their red tail stripes denoted the 500th Bomb Squadron with the Air Apache insignia of the 345th Bomb Group made famous in the Pacific during World War II. The first B-57Bs were issued to this Group. (USAF)

This rare picture of a 66th Tactical Reconnaissance Wing Canberra shows the identifying unit markings. European based RB-57As generally were but one squadron assigned to a wing having other type aircraft. These European based squadrons were disbanded soon after forming and the aircraft returned to the U.S. for assignment mostly to Air National Guard Units. (Buchanan)

essential for special high altitude daytime photography over Hungary during the 1956 revolt. It was inferred that during this time while flying over Eastern Europe, one of the ten modified aircraft was shot down by a ground-to-air missile.

Little if anything has ever been printed about the RB-57A-1 version, probably because of secrecy at the time due to its increased mission capability. When these remaining nine aircraft ultimately reached the 154th TRS at Little Rock, Arkansas in 1961, Roger F. Taylor, a Master Sergeant at the time, was there to incorporate them into his unit. He was of the opinion that the -1s arrived from Europe and not Japan, yet numerous reassignments could account for this. He clearly recalls that as weight savings measures, they were without the heavy bomb bay door and that area was skinned over. The system operator/navigator seat was removed and the night photo flare racks were eliminated from the wings. These modifications had a profound effect on reaching higher altitudes. According to Taylor, to enable the pilot to perform all the reconnaissance duties without the assistance of the photo navigator, an optical viewfinder was installed in the nose of the plane. The clear plexiglass nose cone was replaced with an opaque fiberglass cone. This cone incorporated a small optical glass window for the viewfinder which allowed the pilot to see the terrain from about 30 degrees aft of vertical to about 15 degrees above the horizon. The viewfinder had an electrically rotatable reticle that gave the pilot drift indication as well as camera angle coverage.

Two reconnaissance units in Europe were equipped with RB-57As; the 10th TRW and the 66th TRW. This Canberra of the 66th, 52-1467, has 4-star studded tail and tip tank flash, and yellow fuselage bands. (S. Nicolaou)

The thick wing of the Canberra offered another option to this reconnaissance model when assigned to the 154th TRS. MSgt Taylor developed an installation for all the RB-57A-1s by utilizing the area that formerly contained flare racks of the left wing in which to install a P-2 camera system. One of two cameras was a forward oblique, depressed from horizontal at 30 degrees, with the option of a 45 degree depression. The other camera was a left oblique, with options for 15, 30 and 45 degree depressions, as well as the capability of that camera to be used vertically. Taylor received the Air Force Commendation Medal for devising this installation, the first recipient of this type in the Air National Guard.

Other RB-57As crossed the Pacific and took up residence in Japan with the 6021st TRS for a year and a half, and later in the 6091st both located at Yokota AB. Among these airplanes were RB-57A-1s transferred from service in Europe. Also at the time was one special RB-57A, known by its code name "Switch Blade." Originally based at Rhine Main AB for reconnaissance along the Iron Curtain, it was flown to Yokota by its assigned pilot Jack Reedy for a look-see into North Korea and China. In one side of its fuselage was a roll-top desk type door as a port for the camera, an opening hardly detectable on the ground. This special camera had a 240" focal length and was built by Boston University. Able to see 300 miles if atmospheric conditions permitted, it produced a 9"x 18" format. This airplane and its camera was so closely monitored that its tail number and code was often changed. While at Yokota, 52-1459 being its true serial, it usually had a green "X" on its tail. It was also known to have been re-marked as 52-1618 by November 1963, 52-1423 in 1964, 52-1423 with a green "M", and 52-1421 again having its "X." These 6091st TRS Canberras played a vital role at a critical time in the Far East, eventually being augmented with RB-57Ds.

Air Force Canberras participated in many joint international exercises. A Sabre of the RCAF (or RAAF) flies wing on this RB-57A specially marked with yellow fuselage bands and tail flash for the exercise. (N. Taylor)

B-57Bs are prepared at the Martin plant, Baltimore, for their ferry flights to using organizations. Normally they were first flown to Warner-Robins AFB, Georgia where they were received by the Air Materiel Command there before being further assigned to bomber units. (USAF)

The 461st Bomb Group was the first to be fully equipped with the B-57B bomber version. They demonstrated mobility by two extensive unit movements throughout Central America and Europe. This view, taken at their home base, Hill AFB, Utah, shows a B-57B of the 766th Bomb Squadron with a white meteor as their color. (USAF)

The Michigan ANG received RB-57As at a very early time, their first coming direct from the factory beginning in May 1954. Those that followed came from resources from Germany, for those two units seemed short lived.

While Tactical Bomb Wings and Groups waited for the B-57B Night Intruder models to be completed and delivered, the 345th TBG became the first of the four Bombardment units to receive the Canberra. This was in the form of no less than five RB-57As, however, arriving singly at Langley AFB between June 22 and July 14, 1954. The 345th was not so designated until July 19, 1954, and therefore these assignments were to the 424th and 4400th Bombardment Squadrons of the 4400th Combat Crew Training Group at Langley, mainly for the purpose of proficiency training for maintenance crews. A few of the pilots were checked out before the airplanes were reassigned to the 363rd TRW at Shaw, or returned to the factory. The 345th became the first of the four tactical bomb groups to receive the B-57B bomber versions. It was the 461st Bomb Group (Tactical) that was the first to be fully equipped with the new bombers. After receiving their first B-57Bs in January 1955, at Hill AFB, Ogden, Utah, this unit moved to its new home at Blytheville AFB, Arkansas, over a period from July 1955 to April 1956. As part of the restructuring of the Air Force to 95 wings, the 38th Bomb Group (Tactical) was again activated on 1 January 1953 at Laon, France, with B-26s until these could be replaced with B-57Bs to support the European theater.

In the Far East, the 3rd Bomb Group (Tactical), which served continually in that theater from the beginning of World War II, turned in their B-26s for B-57s, and completed the four group build-up of Canberra night intruder bombers by 1957.

This in-flight view of a B-57B belonging to the 822nd Bomb Squadron, 38th Bomb Group, was taken some place over western Europe in November 1955. This was the only Bomb Group in Europe to be equipped with B-57B Canberras. (P. Pitt)

This flight of four B-57Bs is the aerial demonstration team called the 'Black Knights.' They were part of the 38th Bomb Group stationed at Laon, France. Precision aerial maneuvers performed with these bombers were thought not possible by many onlookers. They performed over many parts of Europe. (D. Menard)

This red meteor followed by yellow and white of the other two squadrons identified the 764th BS of the 461st BG. A yellow meteor followed by white and red identified the 765th BS.

The 3510th Combat Crew Training Wing at Randolph AFB, Texas, provided aircrew training for overseas units beginning in late 1954. Their aircraft were distinctive only by large ship numbers painted on their tip tanks. The 461st at Hill AFB, Utah, also utilized this training while the 345th trained their own crews at Langley AFB. (USAF)

To provide qualified pilots for these units to be re-equipped with B-57s, a transition school was formed at Randolph AFB, Texas, as part of the 3510th Combat Crew Training Wing. Many of the first B-57Bs and B-57C dual control models were initially assigned there beginning in November 1954. To be qualified for B-57 training, pilots needed 1,000 hours total time with twin-engine experience. At Randolph, 25 hours in the T-33 was part of the program to become jet qualified before transitioning to the B-57.

The 3rd Bomb Group in Japan and the 38th Bomb Group in France, each sent three groups of six pilots each for this training. Upon completion, they were to ferry new B-57s across the water to their home units. Pilots from the 461st Bomb Group at Hill AFB received this training also, while the 345th Bomb Group handled their own training at Langley. The training at Randolph, however, did not go as initially planned. Each class was to last for two months, but because of the frequent groundings of the B-57, these classes lasted anywhere from 6 to 13 months. This coupled with delayed production caused many graduates to be returned to their home stations, only to be sent back later when their aircraft were ready for pick-up.

Warner-Robins AFB, Georgia, was the pick-up point for the new airplanes awaiting overseas delivery. Aircraft were usually ferried in groups of four to six. For those crossing the Atlantic, crews from the gaining squadrons augmented ferry crews from the 1737th

On 9 January 1956, the first B-57s arrived at Johnson AB, Japan, for the 3rd Bomb Group. Their former B-26 crews were sent from Japan to Randolph AFB for training, then to ferry their new bombers to their home station in Japan. Frequent groundings of the Canberra in these early days heavily disrupted this scheduling.

Delivering the Canberras to the 3rd Bomb Wing at Johnson AB, Japan was a critical crossing for these airplanes. Departing McClellan AFB, California, for Hawaii, were flights over five hours with ferry tanks, and close to maximum endurance. Crossings were done in groups of four to five airplanes. Johnson AB near Tokyo is now Iruma AB, the HQ airfield for JASDF.

A new type of warrior arrives in Japan. Dressed in the traditional 'protective clothing' of the Samurai, a Japanese host greets the arrival of bomber pilot Ellis Bruch and the Canberra for the 3rd Bomb Group in 1957. Navigator Floyd Pond looks on from rear seat. D. Beggerly

Ferry Squadron at Dover AFB, Delaware. Routing to Europe was through Goose Bay, Labrador; Reykjavik, Iceland; Preswick, Scotland; on to Laon AB, France (B-57Bs) or Spangdahlem AB, Germany (RB-57As).

For aircraft going to Japan, newly trained crews from the 3rd Bomb Group had a lead pilot from the 4440th Ferry Squadron. After a local acceptance flight at Warner-Robins, the new airplanes were flown non-stop to McClellan AFB, Sacramento, California, so that a fuel consumption curve could be plotted before launching out over the Pacific. The crossing was critical due to limited fuel reserve on the longest leg from McClellan to Hickam AFB, Hawaii. Winds had to be just right, with no more than 40kts head wind component for the more than 5hr 30mins flight to Hawaii. An additional 558gal were carried in a ferry tank mounted on the bomb bay door, but more than one B-57 flamed out while taxying off the runway at Hawaii.

The one aircraft that was lost occurred during the initial movement of 56 Canberras to Japan, and that was 916 on 12 May 1957. At Ocean Station *November*, a Coast Guard weather ship mid-point between California and Hawaii, it was learned that Pete Cotellesse's airplane in Black Crow Delta flight of three had a fuel system malfunction which left insufficient fuel to reach land in either direction. The only recourse was for Pete and navigator Gayle P. Johnson to eject near the ship. On the first pass, Gayle was to go out, and on the second pass, Pete would go. Signals got crossed someway however. When Pete told Gayle to jettison the canopy (intending only to clear the cockpit of debris) he took this as the signal to 'go' which is normal procedure. He 'went out' – eight miles short of the ship! Gayle was in the water for nearly 40 minutes before rescuers located and got him on board, but the incident ended safely for both.

The air route across the Pacific after Hawaii was Johnston Atoll, Kwajalein, Guam, then direct to Johnson Air Base, Japan (now Iruma AB, near Tokyo). Each stop along the way was filled with its own and unusual experiences. One hairy incident stands out. Five B-57s were landing at Johnston Atoll, which by its size looked like an

Distinctive markings for Canberras assigned to the 3rd Bomb Wing in Japan consisted mainly of the Roman III on the tip tank with colors used to denote the different squadrons. This yellow marked B-57B belonged to the 8th Bomb Squadron. Fagen via D. Menard

Instrument panel of the B-57 was conveniently arranged. Flight instruments were grouped at center and left, and engine gauges were at right. Fuel control panel was laid out in schematic format, with engine fuel shut-off wafer switches already converted to toggle switches but not shown safety-wired to 'on' position in this picture. Stabilizer trim indicator above canopy switch at left. (D. Beggerly)

aircraft carrier in the Pacific. As number two aircraft was making his 'short field' approach to land, gusty winds at the shore line caused his left wing tank to drag on the coral overrun. This swung the airplane to the left nearly 45deg off the runway heading, directly toward the narrow passage between the fuel storage tanks and the control tower. As all others watched in dismay, the pilot, Bobby Presley, applied full power to the engines that were resting at idle for the landing. The left engine came up to speed first and violently swung the airplane back to the right. Just then the right engine surged to full power and stopped the turn and levelled the airplane on a near original heading. With the gyrations subsiding, Presley raised the gear and gingerly held 160kts as the airplane gradually climbed in a nose low attitude since the flaps were still in the down position. 'Get your flaps up Bobby! Get your flaps up!', yelled I. H. Young over the radio as Bobby passed the end of the runway. His reply, which expressed the tenseness of the situation and afterwards became a classic slogan within the group, was: 'everything is going so good *now*, I hate to change a thing!'

A rash of seemingly unexplained accidents plagued the early days of the B-57s which caused long and frequent periods of groundings. Crashes were far too frequent and early problems with the Canberra took the lives of many crew members. The most common and always fatal accident occurred when flying at high speed at a low level and the airplane would suddenly pitch-down into the ground. Runaway trim seemed the cause, but the reason could never be positively determined. As accidents persisted, with what seemed like increasing frequency, all tactically assigned B-57s were grounded again in May 1956, this time for a period of four months, one of the longest groundings of any Air Force airplane up to that time. One or more of nearly a dozen 'fixes' obviously corrected the pitch-down problem and no further accidents of this nature occurred.

The 'fix' most reassuring to B-57 pilots was that should a runaway nose down trim condition occur at maximum speed, 80lb of pull force – which was well within the bounds of any pilot striving for survival – could hold the nose up until speed was reduced to ease control pressure. Should the trim go to full nose up, this be-

This original B-57B cockpit configuration shows dive brake toggle switch under canopy ledge (arrowed), and stabilizer trim gauge on lower right side panel. Emergency hydraulic hand pump handle was normally stowed. This thruster type ejection seat was replaced in early 1970s. (USAF)

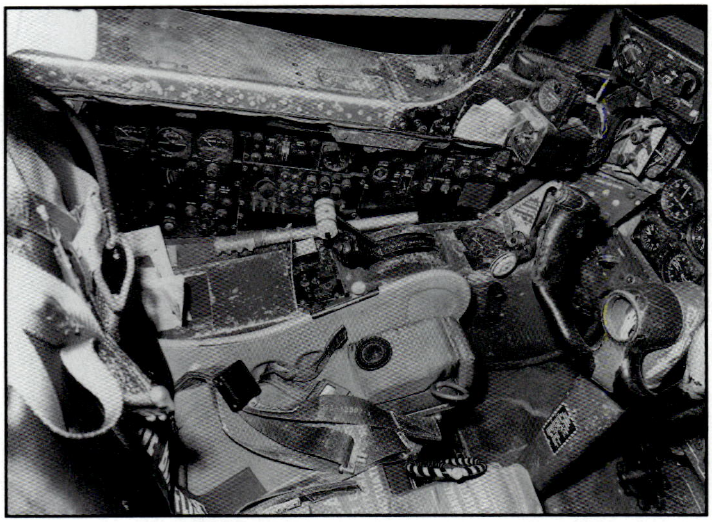

Most noticeable change over the years for the B-57 cockpit was the introduction of the Douglas ESAPAC rocket ejection seat. It could take the seat high enough for ground level escape provided there is sufficient forward speed for chute deployment. A heavy blade knife designed for the nearly impossible task of cutting through the canopy was on a holder below the windscreen.

came uncontrollable above 350kts, but the resultant climb and a power reduction would automatically reduce the speed to where it became controllable. The stabilizer trim gauge was moved to a conspicuous location where it could be monitored easily, and a switch was nearby to cut power to the trim motor.

As with any new airplane, there were many changes to be expected, and some may have already been forgotten – like cigarette ash trays installed in the cockpits when delivered! One of the earliest modifications was the repositioning of the dive brake switch which was hidden under the canopy rail and had to be held throughout the full travel of the 'boards'. This was soon changed to a two position switch easily actuated by the thumb on the right throttle. The canopy open-close switch at the left of the windscreen went through a number of changes from a push-pull to an up-down shielded switch. Few canopies came open and *off* in flight after that change.

The fuel control panel was simple but initially its simplicity created trouble. It resembled a schematic of the fuel system and positioning the control knobs set up the desired flow. However, all the knobs were of the same shape and size – including the engine fuel shut off and bypass switch. When different paint colors also failed to eliminate turning the wrong knobs, these two functions were changed to safety wired switches and a square pull-knob respectively.

We felt that the Canberra noticeably lacked adequate navigation aids in that they were initially equipped with only a radio compass. TACANs were finally added by 1959 and other improvements such as solid state UHF radios, IFF with altitude readout, VOR-ILS, etc, followed over the years. What we called the 'poor man's rudder boost' became installed in the 'Bs' and 'Cs' in 1960. These were not as effective as the full-time power rudder system manufactured into the 'Es', but when the pilot applied heavy rudder force as needed in asymmetrical power situations (single engine), hydraulic power assist supplemented pilot effort. This did not reduce safe single engine airspeed below 155kts however.

The FAA acquired two RB-57As for use in flight testing the high altitude jet route structure. N96 was formerly 52-1438, while N97 was 52-1447. This FAA Canberra N96 crashed in 1960 when the pilot had to eject when the control column inadvertently disengaged as part of the emergency ejection system.

Three B-57s of the 3rd BW in Japan make a formation turn above the Kanto Plains. When the 3rd assumed the nuclear strike mission in the Far East, their 70-83lbs of black paint that had not held up well in the first place, gave way to unpainted, more heat reflective natural metal skin.

Mount Fuji was always a favorite background for flying units stationed in Japan. In this view is 'Green Echo,' one of the B-57C dual trainers assigned to the Training Flight of the 3rd Bomb Wing.

As late in the life of the B-57 as 1972 to 1974, the original ballistic ejection seats with arm rests and actuating grip handles were exchanged for the Douglas ESCAPAC zero-zero rocket seat actuated by pulling a ring positioned between the legs or pulling a curtain down over the face. (These were included in the B-57G modifications in 1969.)

The most unusual handling feature of the British Canberra and early B-57s was that when lowering the landing flaps, the nose would come up, requiring push force on the control wheel instead of neutral or a slight back pressure to hold the nose up. This was soon corrected mechanically by what was called a 'bungee' – an electrically operated system that would apply 25lb of push force to the elevator control system when the flaps started to the down position. This artificial system in this modification made flap extension feel like that of most other airplanes.

This crisp shot of the nose of B-57B 53-3867 shows many details of the freshly cleared skin of the factory applied black paint in 1959. The blue nose indicates that this Canberra belonged to the 90th Bomb Squadron of the 3rd Bomb Wing, Japan.

The control column automatically stowed as a part of the ejection sequence. This cleared the way for the pilot's knees as the seat carried him out of the airplane. I only know of one instance where the column inadvertently stowed, and it took some doing to talk the FAA pilot into leaving the airplane instead of trying to land it with the use of elevator trim alone.

There were all kinds of problems in the early days of the B-57 that caused in-flight emergencies: nose wheels came of, fuel tanks would not feed, causing extremely unbalanced conditions, and many more. In time however, these problems were corrected, and with experience we were better able to cope with unusual situations. The airplane soon lived up to its full expectations.

By the time the four bomb groups received their full inventory of new B-57s, a number of tactical developments and exercises were geared to these new jet bomber units. The first of these was Exercise Sagebrush in November 1956, which took place across the lower eastern portion of the US and involved both US Army and Air Force units. The 461st TBG and 363rd TRW were the ag-

OVERLEAF: This is a typical ramp view of 3rd Bomb Wing B-57 aircraft at Yokota AB, Japan around 1964. The 40th FIS and its alert hangars with F-102s is in the top of the picture at the south end of the base. The 3rd BW departed soon after this picture was taken, influenced by the Vietnam War.

U.S. AIR FO

For the night intuder mission profile, an overall gloss black finish with red lettering was adopted in the late 1940s, as illustrated by this B-57B of the 82nd BS, 38th BG(T), based at Laon in France as part of USAFE. Replacing the group's veteran Douglas B-26s, the B-57Bs with their flamboyant unit markings were brief residents in Europe, with aircraft beginning to be ferried back to the USA in 1958.

Showing pride in their unit and their newly acquired B-57s, members of the 822nd Bomb Squadron, 38th Bomb Wing at Laon AB, France, pose for a formal picture in 1956. (D. Menard)

gressors and demonstrated the swift striking capability of jet strike force by achieving a quick victory. In 1957, 13 B-57s of the 461st made up Task Force 'Vista Able' for a goodwill flight to several Latin American countries extending as far south as Tulara, Peru. Again this same wing took part in 'Mobile Charlie', a deployment to support Exercise Counterpunch in Europe in 1957, aided by B-57 Canberras of the 38th Bomb Group at Laon AB, France.

It was on the return to the United States from 'Mobile Charlie' in Europe on 6 September that the 461st had a very close call. At the point of no return out of Keflavik AB, Iceland en route to Goose Bay, Labrador, 1Lt James E. Kater recognized a fuel system problem (in 53-3885) which made the wing tank fuel unusable.

Flight leader Capt Albert T. Keeler (in 53-3934) took the two ships to 45,000ft where lighter winds were reported. Keeler's navigator, Robert B. McMullen, who was the 764th Squadron navigator, coordinated with 1Lt Don H. Hall, Kater's navigator about the problem. McMullen gave Keeler assurance that Kater and Hall would probably be flamed out at the destination, but they could make it, establishing positive thinking in the flight. Hall calculated they would only have 600lb of usable fuel upon reaching land fall, but would have another 127nm to go to reach Goose Bay AB against 100 to 120kt headwinds. The flight was over the southern tip of Greenland, and to save fuel, Kater jettisoned his dry tip tanks. Air Sea Rescue was altered and a SA-16 Albatross followed the aircraft toward Goose Bay. When radar reported landfall, Kater shut one engine down to save fuel, placed the other at idle, and began his descent toward Goose through some cloud layers. Keeler hung on through the clouds, staying on Kater's wing to help work through the problem. This was not a new experience to Keeler, for two years before in May 1955 he made the first dual flame-out emergency landing for a B-57 at Scott AFB, Illinois. For this display of airmanship, he was awarded the Air Medal.

At 8,000ft altitude, GCA gave their position as 2 miles out. The F-89 positioned at 'high key', which Keeler had requested to give field location from above the undercast, reported that he was directly over the field. After Kater put the gear down, while still

LEFT: Courtesy of Aerospace Publishing, Ltd.

A change in mission for the B-57 from night interdiction to delivery of special weapons called for the removal of heat absorbing black paint. These 345th BG Canberras start engines for a formation training mission at Langley AFB, Virginia in 1957. (M. Olmstead via D. Menard)

Canberras of the 3rd Bomb Wing at Johnson AB, Japan, were easily recognized by the Roman numeral III on their tip tanks. The yellow markings on 53-3894 signifies the 8th Bomb Squadron. (Fagen via D. Menard)

Cross country flights were normal training missions for B-57 crews. Here 52-1559 from the 461st Bomb Wing refuels at Shaw AFB in May 1955. (A. Keeler)

B-57s were often placed on exhibit for static display functions. Sent from the 461st Bomb Wing, 53-3934 is shown here while on exhibit at Andrews AFB, Washington, D.C. in 1956. (L. Davis)

Armed Forces Day At Rhine Main AB, Germany included a Canberra from the 822nd Bomb Squadron, 38th Bomb Group. This is the often pictured 52-1577 "M" when photographed in 1957. (S. Nicolaou)

Mito Range was the frequent practice target for B-57s of the 3rd BW in Japan. This rare view shows 879 having just released a 28lb practice bomb in a glide bomb delivery from its spacious bomb bay. Rocket and gun passes, as well as skip bomb, LABS and Shoran drops were also practiced regularly. (R. Barnett)

A full generation of pilots flew USAF B-57s. Barbara Lynn Bruch, age 10, welcomes her daddy back to Japan from Korean deployment in 1958. Eighteen years later she married B-57 pilot, 1Lt Timothy Killeen, a member of her father's unit, the 4677th DSES while at Hill AFB, Utah. (E. Bruch)

having power with which to do so, the two aircraft began a 360 degree gliding turn to lose about 4,000ft of altitude. During this descending turn, Kater and Hall lost their last engine to fuel starvation. As they rolled out of the circle, they broke through the bottom of the clouds at 4,000ft, with the landing runway directly beneath them! This was a perfect position to begin a second 360 degree turn, gauging every degree of bank for the touch-down point. Kater frantically hand pumped the flaps and dive brakes down to kill off excess altitude while turning base and final. With this double engine flame out, he touched down at about the 2,000ft marker. Pilot skill coupled with the teamwork displayed by Air Rescue Service, Goose RAPCON, element leader Al Keeler and other Goose agencies were directly responsible for the *save* of a valuable aircraft and possibly the lives of two crew members. Kater and Hall were awarded the Air Medal for this action.

On 2 April 1964, the 3rd Bomb Wing no longer had a nuclear strike commitment and returned all its forces to its home at Yokota AB, Japan. The return of 20 airplanes filled the ramp to overflowing with a full wing complement of B-57s. Wives and friends met the B-57 crews at their planes for this highly celebrated event.

Above: This very rare picture was taken of 3rd Bomb Wing aircraft on alert within the secure area of Pad C at Kunsan, Korea. Since atomic weapons could not be positioned in Japan, immediate strike force B-57s had to be kept in Korea. One 17ft long Mk7 special weapon is loaded on each of these 'Quick Strike' Canberras. Below: A detail of the above photo. These Canberras have their hatches open for their daily morning inspection by the bomber crews while standing alert duty at K-8, Kunsan, Korea. These B-57Bs were loaded with Mk7 special weapons and in readiness for the first three bombers to be launched in 15-minutes after the signal was given. This picture of Pad C was taken in early 1964.

The 'Pickle Barrel' trophy was the focal point of esprit de corps for the 3rd Bomb Wing, being awarded to the squadron having the highest overall B-57 weapon delivery rating during a quarterly period. As the 3rd BW deactivated in 1964, Lt Col Fred Grindle, CO of the 8th Bomb Squadron, accepts again the coveted award for his squadron. To his left are Carl Bratten, and toasting is 'Bear' Barnett, both killed in the 8th BS after it deployed to SEA. At right is Operations Officer Howard O'Neil, destined to drop the first live ordnance on an enemy from a USAF jet bomber while CO of the 13th TBS. To his right is author Bob Mikesh, pilot of winning crew of the 8th. The 'Pickle Barrel' stayed with B-57 units throughout the war in SEA and was retired to the Air Force Museum from Ubon when the 13th TBS was deactivated in 1972.

After three short years with these B-57s in tactical bomb groups, the units were programmed to be phased out. The 38th was the first to begin by ferrying their aircraft back to the U.S. in early 1958. Soon afterwards, as of 1 April 1958, the 461st at Blytheville was also deactivated. No sooner had the B-57 bomber force been removed from Europe, however, than a threat to peace developed in Lebanon. To provide a show of force, B-57s from the 345th Bomb Group at Langley were deployed within a three hour notice in July 1958, for Turkey, as part of Composite Air Strike Force 'Bravo.' Once in place at Incirlik Air Base, Turkey, nothing happened, but they remained there over three months, and ready for any action.

On the other side of the world, another crisis erupted in the Taiwan Straits. Again the 345th was called upon for support, and 12 or more B-57s were deployed to Okinawa on 29 August 1958. Although B-57s of the 3rd Bomb Wing were close at hand in Japan, world tension kept them tied to their assigned strategic targets in that part of the world. For this time period, the strength of two squadrons of Canberras from the same 345th Bomb Group in the U.S. were sent off in opposite directions on tactical operations and nearly met on the other side of the world. When the dust settled, the 345th was disbanded on 25 June 1959, leaving only the 3rd Bomb Wing to survive for nearly another five years as the only tactical bomber wing within the USAF. Its existence seemed essential, however, as its primary mission was a SIOP (Single Integrated Operations Plan) commitment for 'Quick Strikes' against strategic targets on the mainland of China, North Korea and Russia. Since nuclear weapons could not be maintained in Japan, the 3rd Bomb Wing set up a rotation of aircrews to stand alert at Kunsan, (K-8) Korea, with nuclear armed B-57s which were ready to be launched against preplanned targets, and to be airborne within 15min notice. At first, alerts lasted for a month at a time as each of the three squadrons took a turn, but soon changed to a two week crew by crew rotation. This continued for an agonizing period from August 1958 to 2 April 1964. Approximately one-third of the wing was at Kunsan all the time, which meant for the air and ground crews, one-third of their time was spent away from their families living in Japan. I know, for I had my share of deployments to 'Pad C' at K-8 for the last year of this period.

8

Clark AFB and the Bien Hoa Deployment

This new chapter for the B-57s began with a plan that would inactivate the 3rd Bomb Wing and bring to an end the last remaining Tactical Bombardment Wing in the United States Air Force. As the wing and its squadrons were to phase down, crews being rotated to the United States would ferry their Canberras to the Massachusetts, and New York Air National Guard. Before 2 April 1964, which was the last day of the 'Quick Strike' alert commitment at Kunsan, Korea, five B-57s (918, 935, 897, 900 and 854) in two flights had already departed Yokota Air Base. The escalating war situation in Southeast Asia changed the inactivation plan in mid stream and sent the 8th and 13th Bomb Squadrons to Clark Air Base in the Philippines, and returned the 3rd Bomb Wing and 90th Bomb Squadron on paper to the United States. On 9 April 1964, 12 of the first Canberras arrived at Clark, led by Lt Col Billy A. McLeod, Squadron Commander of the 13th. The 8th began arriving on 17 April led by Squadron CO, Lt Col Frederick W. Grindle,

As a salute to their new home, 12 B-57s of the 13th Bomb Squadron arrive from Yokota AB, making a formation fly-by upon arrival at Clark AB, Philippines, 9 April 1964. This sudden reassignment was initially a temporary one, dependent upon the war situation in the theater. (I. P. Graham)

With Mt Arayat prominently across the field at Clark AB, a B-57 from the 8th BS, flown by Roy White and John Kendrick, touches down on the runway at their new home base. The strength comprised 47 Canberras outfitting the 8th and 13th Bomb Squadrons.

Somewhat traditional when a squadron arrives at a new home station, the 'red carpet' is rolled out for the lead crew. Lt Col Grindle, CO of the 8th BS is about to receive royal treatment on his arrival with his squadron at Clark AB. (I.P. Graham)

Committed to the combat zone in August 1964 a detachment of B-57s from Clark stand ready on the east ramp at Bien Hoa AB near Saigon. Conditions were crowded, and protection from enemy assaults was minimal. (USAF)

Jr, and by the 29th, all 47 of the former 3rd Bomb Wing's B-57s were in place at Clark.

A war of nerves for the B-57 crews continued to build as the imminent war situation in Vietnam escalated. Although US forces were present in Southeast Asia at the invitation of the South Vietnamese government as military advisors, direct military force could not be applied unless provoked by North Vietnamese aggression. The provocation needed, however, occurred on the night of 2 August 1964, with what is historically referred to as the 'Tonkin Gulf Incident.' North Vietnamese torpedo boats made deliberate attacks upon the U.S. destroyer *Maddox* while it was on patrol off the North Vietnamese coast. This, and a subsequent attack two nights later, touched off an exchange that brought the United States into direct military involvement with Viet Cong and North Vietnamese forces.

At Clark, everything was in a wartime atmosphere. Fighter units were deploying, air transports were lifting supplies, yet the two B-57 squadrons were held on ready standby. Should they move, their destinations would be either Takhli Air Base, Thailand, or Bien Hoa (pronounced Ben Wa) – an airfield about 10 miles north of Saigon. Anxiety by the B-57 crews grew tense, knowing it was going to happen – but *when*?

An alert notice to deploy the two B-57 squadrons came just before midnight on 4 August, yet by early morning disappointed crews were told to return to their quarters to again standby for further information. They were anything but relaxed. By that evening, the movement order came by message ordering deployment of 20 B-57s to Bien Hoa Air Base as soon as possible. Crews had already been assigned to aircraft, and now flight plans to this just announced destination were hastily prepared for the five flights of four aircraft each. Lt Col Billy McLeod led the first flight off at 19:00hrs and each succeeding flight launched at 20-30min intervals. What had not been taken into consideration as the hours of the day dragged by without a word, was that the deployment might be made at night. Few of the crews were current in night formation as it had not been a B-57 combat crew training requirement. Regardless, the departure in flights of four worked out quite well for their first time in darkness. It was at the other end where problems soon developed. Bien Hoa was an unfamiliar field for most of the crews, especially with added restrictions caused by darkness and the 700ft ceiling of monsoon weather. Uncertainties after landing caused one aircraft to delay on the runway, allowing another with loss of hydraulic brake pressure to overtake it in the darkness and collide with it. As a result, 884 was a write-off, however, after much work, 877 was repaired. Adding even more to the problem, another B-57 blew both tires on landing roll which further tied up the runway. Fortunately, there were no injuries from these mishaps.

One last check of the 500lb low drag bombs on the B-57 wing pylons is made by Enos Chabot of the 8th BS. Initially B-57s were not involved in combat strikes for the war was restricted to propeller driven airplanes only, this being established by the Geneva Accord of 1954. (USAF)

Maintaining a show of force posture in South Vietnam a formation of B-57s start all engines simultaneously at Bien Hoa for departure on assigned road reconnaissance routes. This became painfully dull for the crews who knew that propeller driven aircraft were allowed to make air strikes against the enemy in accordance with the Geneva Convention guidelines. (USAF)

In the early morning of 1 November 1964, the east ramp at Bien Hoa AB was aglow with burning B-57s after a mortar attack by the Viet Cong. Five Canberras were destroyed and many others damaged. (USAF)

The remaining B-57s destined for Bien Hoa had to be diverted to Tan Son Nhut, a few miles away. Deviations from intended flight plans while en route can become the basis for a bad situation, at night, and with so many aircraft involved. For reasons which were never satisfactorily determined, Capt Fred Cutrer, and his navigator, Lt Kaster crashed in 870 a few miles short of the runway at Tan Son Nhut while on a straight-in TACAN approach. This crash site was in a heavily populated VC area and was considered a combat loss. It took weeks before a diversionary and a sizable main force could be staged to retrieve their bodies from the crashed plane.

These losses made anything but a good showing for the B-57s at the start of their operations in Southeast Asia. Investigation of these mishaps traced the root of the cause to the movement order and the interpreted meaning of 'as soon as possible.' In a wartime situation, most commanders would construe this to mean 'now', while in a peacetime atmosphere, it would imply to move after proper crew rest, and under suitable flying conditions commensurate with training received. Needless to say, later messages of this type were given closer scrutiny before issue following this deployment.

A fighting force of B-57s was now in the combat zone and ready for action, however it would be seven months before the Canberras would be given an active role in combat. In the meantime, their presence in Vietnam displayed to the invaders a stronger US fighting force, which may, or may not, have had an effect on enemy strategy. Prior to this move and many months after, two or three B-57s were assigned courier duty on crew rotational basis at Tan Son Nhut.

Now that a portion of each of the two bomb squadrons were deployed to South Vietnam, their presence was given little notice in the war effort. Seemingly as a pacifying gesture, they were given road reconnaissance missions but no engagements or attacks were allowed, and the B-57s flew unarmed. They were to cover the roads leading out of Saigon, and report any enemy ground activities or road blocks along these routes. These missions, although flown with much enthusiasm initially, proved to be futile the majority of the time. Occasionally a flight would report a road block thrown up by the Viet Cong, however, these reports seemingly were not utilized in strike missions. Sortie requirements degenerated to four per day, and to keep everyone busy, one crew would fly for 30-45 minutes, then land so that a crew change could be made and continue with the same mission. By this time the Bien Hoa commitment, more formally called 405th ADVON 1, was an unpopular TDY (Temporary Duty) as far as the air crews were concerned. This attitude was due, in part, to the extremely limited activity afforded the B-57 aircraft in the Vietnam conflict. Other air units were launching out of Bien Hoa, and were under a full combat commitment while the Canberras were restricted from actual strike missions. Tensions mounted, caused by the 'ribbing' dealt out from those engaged in the war. To ease this strain on morale, squadron members were rotated regularly so that the TDY period was only one week per month for each crew. Half of the aircraft were returned to Clark on 15 October to reduce the TDY burden and the load on the overcrowded base. This proved fortunate.

It was late on Halloween night 1964, hardly three months after the arrival of the B-57s at Bien Hoa, when another disaster happened. A surprise enemy mortar attack caused a crippling blow to the base and its occupants. Everyone talked of these possibilities, but no one did anything about it. Seemingly, the reason being that the bomber squadrons and other American units were merely ten-

Shrapnel opened the wing tanks of 555 allowing its fuel on the ramp to be ignited by the airplane burning next to it. This airplane was a total loss along with four others. (USAF)

Two Canberras were completely burned to the ground leaving only their jet engines. Had this enemy attack taken place a week earlier, twice the number of bombers were on the ramp and losses would have been greater. (USAF)

Mortar attacks against US bases during the war were a common occurrence. In addition to B-57 losses at Bien Hoa that night, four Americans were killed and 72 others wounded. President Johnson called for immediate replacements for the Canberras to be sent from the U.S. (USAF)

A fork lift truck gathers the few remaining pieces of 894 to clear the ramp at Bien Hoa. The B-57 squadron felt these losses, but they did not let them restrict their operational flying commitments against the enemy.

ants on the base, and base security was that of the ARVN forces. But at 00:25hrs on 1 November, the expected happened. The Viet Cong launched a stand-off mortar attack against the base, using six 81mm mortars emplaced less than a quarter mile outside the northern perimeter. Communist gunners fired between 60 and 80 rounds into the base and swiftly departed, undetected and unmolested. At the sound of the first incoming enemy fire, everyone took cover. The explosions which followed, coupled with the illumination caused by flames made it evident that the rounds had hit home. The Canberra crews were billeted in the American compound located in the center of the main base, and from the direction of the fire seen against the sky, there was little doubt that the B-57 ramp had been hit. Of the 20 Canberras parked wing tip to wing tip in four rows, five were destroyed, of which two were burned to the ground. These were 555, 892, 894, 914 and 924 Four Americans were killed and 72 others wounded. The Viet Cong were the victors that night, inflicting damage out of all proportion to the effort expended. This became the bomber squadrons first losses due to combat, and they had not had a chance even to drop one bomb up to this point. In addition to the five Canberras lost, 15 others were damaged and four VNAF A-1s were destroyed or damaged.

There were few changes after the incident as ramp conditions were such that sandbagging revetments in the cramped space was impractical. To give a smaller target should there be further attacks, the number of B-57s at Bien Hoa was reduced even further.

The next problem at hand was in maintaining trained crews for the Canberra squadrons. The *old heads* that came with the airplanes from Japan were nearing the end of their three to four year overseas commitment. Many new arrivals were without prior B-57 experience and the two squadrons were launching 40 to 50 training flights a day, making Clark resemble a training base. The maintenance workload on the airplanes was drastically increased.

The problem was alleviated somewhat by new pilots receiving B-57 transition training with the Kansas Air National Guard at Hutchinson, prior to leaving the States. President Johnson recognized the need for replacements and had them provided. Eight replacements came from the Kentucky ANG as Project *Red Cherry* at the beginning of 1965, followed by two RB-57Bs from the Nevada ANG. Nine more Canberras came in May of that year from the Nevada and again from the Kentucky ANG. Others trickled in over the next twelve months maintaining the two squadron at acceptable levels.

9

First Blood

Nothing remains stationary in a war situation and in the opening months of 1965, a major change in air tactics was being formulated. Morale of the South Vietnamese was at a very low ebb. There was no appreciable change in the war situation that indicated progress in their cause. American air power was present, including jet fighters, but only propeller driven aircraft were being used. The Americans were abiding by the Geneva Accord of 1954, which stipulated that no new military weaponry would be introduced in Vietnam by either side. Canberras and other jets continued to fly unarmed visual reconnaissance missions. However, in the previous year, the North Vietnamese had violated that agreement and the South Vietnamese Government asked the United States to allow jet planes to be used for combat strikes.

Word reached the Canberra crews that should the restriction be changed, B-57s would be used. On 19 February, that day came, setting a new milestone in the Vietnam war. With Viet Cong troops achieving impressive gains, U.S. Army General Westmoreland, Commander, MACV (U.S. Military Assistance Command, Vietnam) – invoking authority given him three weeks before to use jet aircraft under emergency conditions, ordered the B-57s into action.

Both engines of this B-57 at Bien Hoa come to life simultaneously as the last trace of smoke belches from its starter cartridge exhausts. On the wing stations of 888 are 500lb GPs with an internal load of 21 260lb fragmentation bombs. (USAF)

The first American jet bomber to drop live ordnance against an enemy was this B-57B, 53-3888 of the 13th Bomb Squadron. The attack took place on 19 February 1965 against a suspected concentration of Viet Cong guerilla forces near Bien Gia, east of Saigon. (USAF)

For two days, B-57 crews had been on alert, and when the execution order was issued at 14:00hrs, they were ready. The first flight of four responded with exuberance and were airborne at 14:30 hrs. Twenty minutes later they were over the target, a suspected concentration of Viet Cong guerilla forces near Bien Gia, about 30 miles east of Saigon. A Forward Air Controller (FAC) in an O-1 Bird Dog had received ground fire and marked the target with smoke rockets when the B-57s arrived. Maj Howard F. O'Neal, Commander of the 13th Bomb Squadron, with Maj Frank R. Chandler, Squadron Navigator, peeled off for the attack and became the first crew to deliver live ordnance against an enemy from a US Air Force jet bomber, that Canberra being 53-3888.

Following closely on the heels of this first attack was another B-57 flight of four, led by Capt Bennett Stone of the 8th Bomb Squadron. In 905, with Ray Pence as navigator, Ben made the first bomb drops for the 8th. Through these attacks, several secondary explosions were observed by the crews and the FAC, but the overall damage inflicted on the enemy was obscured by the dense jungle foliage. A total of 18 B-57s attacked that target and all of the Canberras returned to Bien Hoa unharmed. F-100 fighters joined in the attack later that same day. Ironically, a party had been organized that night for those Canberra crews at Clark. The mood skyrocketed when Maj Roy White, Commander of the 8th announced the news of the raid. Morale literally zoomed, and the next day a high spirited augmentation force of three more crews left for Bien Hoa.

Two days after this first attack an Army Special Forces team and a Civilian Irregular Defense Group (CIDG) company were caught in a communist ambush at the Mang Yang pass on Route 19, east of Pleiku. Supported by F-100 and B-57 strikes, which prevented the enemy from overrunning Allied forces, U.S. helicopters moved in and successfully evacuated 220 men who might otherwise have been lost. Although the military situation in Vietnam remained discouraging, the events of February 1965 marked a turning point in the history of the war. Officials in Washington no longer talked about withdrawing American military advisors, but instead, now recommended deployment of additional U.S. forces to Southeast Asia, viewing a swift conclusion to the war. The stops had been pulled for action by the B-57s and when March arrived, they were scheduled to fly their first out-of country mission in a war that seemed to have no territorial boundaries.

To provide maximum support, more planes and crews arrived from Clark bringing the total force to 28 planes and 30 crews. Three possible targets were planned and studied, all seemingly in high risk areas having radar controlled AAA gun protection. The selection became Xom Bang, a military staging area 17 miles north of the DMZ (Demilitarized Zone). Despite one day delay due to weather, the attackers became airborne at 14:50hrs on 2 March, dampened only by a drizzling rain. One aircraft aborted out of 20, and it was replaced by one of two airborne spares. The Canberra crews were leaving nothing to chance for not giving a good showing. Spaced 10 minutes apart, the bomber formations of five flights

USAF
U.S. AIR FORCE

USAF
U.S. AIR FORCE
BA-833

Bombs were the most effective weapons carried by B-57s but, special missions sometimes called for the addition of rockets. Four 7-tube rocket pods were carried on wing stations for 2.75in 'Mighty Mouse' rockets. 'Yellow Quebec' (tail letter) of the 8th BS salvoes on the enemy below.

A Martin B-57 of the 13th Bomb Squadron takes off from Bien Hoa AB, for an air strike against the insurgent Viet Cong forces in Central Vietnam. (USAF)

of four planes each, passed their checkpoint at Da Nang. Here they turned their radios to the target area control frequency, only to be stunned to hear that three F-105 flak-suppression aircraft had already been downed by intense AA fire in the target area. Heavy resistance was expected, and this confirmed it.

The bombers pressed on and as the former capital city of Hue (pronounced Way) came in sight, they went 'on the deck' to avoid radar detection until the pop-up for the actual dive bomb run. All crews were to make one pass only, releasing all ordnance, but a second pass could be made only if the first was not successful. At the time of the arrival of the lead flight on target, parts of the target were already smoking from flak suppression ordnance. The Canberras led by Maj Roy White came in and dropped without difficulty. After his flight had made their attack, they stayed nearby and flew air cover for the other flights until low fuel demanded that they depart and recover at Da Nang. Capt Fred Huber, leader of the last flight also recovered at Da Nang with low fuel, while the others returned safely to Bien Hoa.

Some battle damage was incurred but none serious enough to cause any major problems. The Canberra crews felt they had at last earned their place when bomb damage assessment photos showed over 95% destruction in the target area. The heavy loads carried by the B-57s had indeed proven most deadly against the North Vietnamese. The 20 aircraft dropped over 96 tons of bombs, with a distance to the target of 450 miles.

As the missions became more challenging with enemy resistance increasing, combat losses were soon to begin. The first of these occurred on 11 March 1965. Bill Mattis and Dick Smith were in 890 on a two ship in-country strike, when the Canberra erupted in flames during an attack and crashed about three-quarters of a mile beyond the target. Ironically, Jim Lewis, who was flight leader on that mission, was the next to be lost along with his navigator Art Baker on 7 April. His Canberra 880 failed to recover from a dive-bombing pass, apparently being hit by ground fire during the attack.

PREVIOUS: Markings that were used by the 3rd BW in Japan remained with these B-57s in SEA until they were camouflaged. Here, flying over the Mekong Delta is 'Yellow Romeo,' 879 in the lead. It retains its yellow fuselage band denoting the 8th BS. On its wing is 833 with its red fuselage band of the 13th BS. Since only one of the two squadrons was in country at a time, color assignments were no longer valid. (USAF)

10

Night Intruder Missions

While B-57s proved their worth as excellent close air support aircraft during the day, it soon became evident during this time period that it was the most suited aircraft for the night interdiction role. Enemy supply lines flourished under cover of darkness and in March 1965, B-57 night missions out of Bien Hoa began with a determination to slow this traffic. In anticipation of this, the two B-57, squadrons had begun night flying training at Clark at the turn of the year.

The first night missions were led by Capt Fred Huber of the 13th, over a free strike area 68 miles south of Da Nang. The technique was for the lead aircraft to locate the target area and drop parachute flares for illumination. The second aircraft then made regular ordnance passes with bombs and guns. Later, at the direction of 2nd Air Division, Maj Howard "Howie" O'Neal experimented by using a C-130 as the flare ship which gave freedom for both B-57s to work the target. Night attack training of this type

This was a strange mix of aircraft on a very coordinated night mission attack. Ably portrayed in "Night Intruders" by Keith Ferris in this painting, the C-130 used night sensors to search for ground targets. Once identified, the C-130 dropped flares, and the B-57s made their attacks. The Marine EF-10B kept electronic surveillance over possible enemy missile attacks. (© 1980 K. Ferris)

Like clockwork, another Canberra leaves Bien Hoa to seek out the enemy and unleash its deadly load. This airplane was armed with four M39 20mm cannon with 290 rounds totalling 1,160 for the airplane. Earlier aircraft from 52-1493 thru' 575 were equipped with eight M3 .50cal machine guns of 300 rounds each, totaling 2,400 rounds for each aircraft. (USAF)

continued in-country through the rest of March and into mid-April. By the time the B-57 crews ware trained and ready for night intruder work along the supply routes in Laos, a well coordinated system had been worked out. Although not all night missions were flown in identical fashion, one in particular proved very effective. Two B-57s would depart Bien Hoa climbing to 30,000ft on a route that would take them to a predetermined location, usually a TACAN fix or a prominent geographical location, generally across the Laotian border. Approaching this location, it was not difficult for the lead bomber pilot to sight the rhythmic flash of the rotating red beacon on top of the C-130 flare ship. Called 'Blind Bat', this C-130 would be circling at 15,000ft where a most unusual formation would form. The two B-57s would join on the C-130, one on each side, remaining slightly high so as to maintain position by the white lights displayed only on the top of the aircraft. The red beacon would now be turned off. Below and slightly to the side was a strange friend to this incongruous formation; a Marine EF-10B Douglas Skyknight. This two-place twin-jet straight wing fighter even predated the Canberras, and a squadron was stationed at Da Nang. They were an excellent airplane for the purpose. Protecting this air strike team, they jammed radar controlled AA and detected hot missile sites that might be preparing to launch. One B-57 pilot recalls that a twin-engine Grumman S-2 Tracker filled the space of the EF-10B on at least one occasion. As if this ware not enough, an RB-57E from the 'Patricia Lynn' unit at Tan Son Nhut joined in a trail position behind the C-130. This aircraft equipped for night photography would take real-time photographs to record the night's accomplishments. After the first few weeks of this night operation, the 'Patrica Lynn' accompaniment was discontinued.

The B-57 proved to be the best airplane in SEA for night intruder missions the purpose for which it was selected 14 years before. Good visibility from the cockpit without structural hindrances was a marked feature. It could be slowed in flight sufficiently for the crew to have time to pick out targets and line-up on an attack run. Good loiter time and large bomb loads were also important assets.

Once joined, the formation proceeded to a point on the vital highway network, dubbed the 'Ho Chi Minh Trail' that fed the enemy to the south. The mission was to destroy any truck convoys that might be sighted as they carried supplies on their southern

Many tasks in the war in SEA were performed by the Lockheed C-130 Hercules. One of these was to act as flare ship for B-57s. Known by their call sign 'Blind Bat', two B-57s would fly formation with it to the target, then when it was lighted by flares from the C-130 the Canberras would break away and begin the ground attack. (USAF)

To protect the night formation by detecting hot missile sites, Douglas EF-10Bs from a Marine squadron at Da Nang joined the mixed group of aircraft for night interdiction attacks.

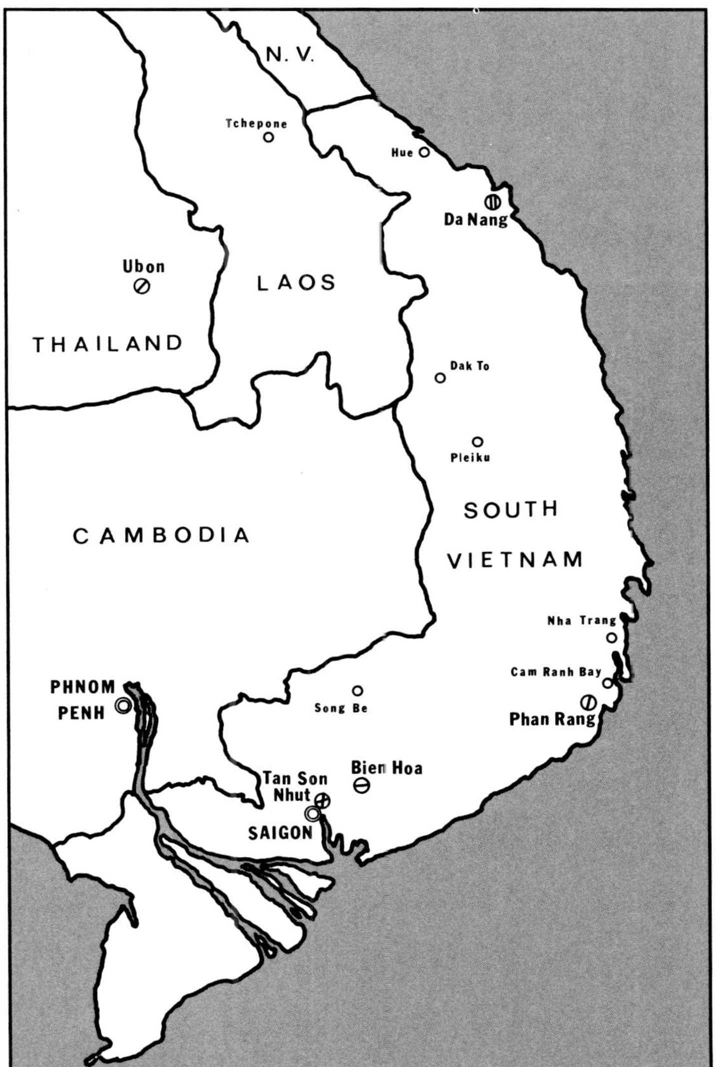

South Vietnam showing bases associated with B-57 operations.

movement. It was difficult to locate trucks visually at night. They moved with their lights blacked out or dimmed, then would pull into the woods when aircraft were heard overhead. The darker the night, however, the easier it was to spot convoys now forced to use lights. With the formation now at 8,000ft, in the mountainous area, it was the crew of the C-130 that would spot the convoy. When located, the C-130 commander would call it out to the other aircraft in the flight and indicate he was preparing to drop flares in two or three minutes. The two B-57s would move off to the side and drop down about 1,000ft below the flare ship. The C-130 pilot would announce when the flares fell away. Twelve evenly timed sparks could be seen as the igniters deployed small parachutes that suspended the flares. Moments later a string of white flares would light up the landscape. The road would become quite visible. But it would be rare if a truck could be seen unless caught in an open unsheltered area. At least their hiding locations along the side of the road in the area they were last seen could be estimated and it was time for the B-57s to do their work in the few minutes the flares lasted. Each Canberra took its turn dive bombing and strafing the target area, calling off their positions by radio so as not to interfere with the other crew's attack.

With the ordnance loads that the aircraft carried, it was the rule rather than the exception that one or more trucks would be hit and explode or burn, even though concealed from sight. Sometimes this would cause a chain reaction with nearby trucks, and the kill for the night could become quite high. Even if missed, the mountain roads would be heavily damaged and required extensive repair which decreased truck traffic considerably – which was the objective.

Most of these strikes would be interrupted by automatic weapons fire from emplacements strategically located along all the supply routes. The same flares lighting the ground also silhouetted the B-57s at certain points during the attack. As Canberra crews will attest, their defensive fire could be effective. The EF-10Bs were able to keep track of the big stuff and call a warning, but close to the ground, the B-57s had to battle it out alone. Many a tail feather came home riddled. Some did not come back at all!

The first of these night missions into Laos took place on 1 April 1965 in the area of Tchepone, where the supply routes filter in from North Vietnam. Capt Ben Stone of the 8th Bomb Squadron

Final check of 500lb GPs on the bomb bay door is made by the navigator of this B-57 just prior to take-off. While the low stance of the Canberra eased most maintenance tasks, its low bomb bay gave little working room for loading and inspecting bombs. (USAF)

Armorers attach 750lb napalm tanks to the outboard wing stations of this B-57. Technicians are Gene Allard, Bill Benson, Bob Walker and Delmar Thaden. Napalm was an effective weapon against area targets such as Viet Cong supply build-ups, helping to keep the enemy on the move and off balance as much as possible. (USAF)

led the flight of two B-57s and joined with the C-130. As flares began illuminating the area of a suspected river crossing, Bev McCord, the navigator in the back seat spotted a ferry boat and the attack began. Art Jepson and Dick Ryan of the 13th in the second ship got near misses followed again by Ben Stone, who drove in the sinking blow. When the strike ended, the estimate of damage was one ferry boat destroyed, four trucks sunk with it, one bridge and approaches damaged and approximately 20 buildings destroyed by secondary explosions. Pressing the mission even further, they continued northbound to the Mu Gia Pass on the North Vietnam border. This first night mission into Laos marked the beginning of Operation Steel Tiger; a limited day and night US air campaign against enemy troop and supply movements within the panhandle of southern Laos. It joined with a strictly daytime operation called 'Barrel Roll', which had begun on 3 March aimed against Communist Patho Lao and North Vietnamese troop movements. By December 1965, 'Tiger Hound' would become another familiar term well remembered by B-57 crews for strikes along the vast network of the Ho Chi Minh Trail, all identifying areas of interdiction missions of enemy supply routes.

A dolly of M-117s; 750lb GPs stand ready to be loaded aboard these B-57s at Da Nang in May 1966. The Canberra's bomb bay and wing racks can carry a variety of bombs. Space, not load carrying ability, was often the determining factor for these numbers. (USAF)

Eventually these night interdiction missions were extended into North Vietnam, and the first of these took place on 21 April 1965. The mission got underway when Capts Fred Huber and Howard Greene got their Canberras airborne in this flight of two. Ninety minutes later they were in the target area, a vast spider web of supply routes leading south from the city of Vinh, mid-point between Hanoi and the DMZ. At this time, the C-130 flare ships had not been approved for North Vietnam missions, so each B-57 carried MK-24 flares as well as six 500lb bombs and eight Mighty Mouse rocket pods containing seven 2.75in rockets each. Attacks were made against a wooden railroad bridge and a concrete span bridge just north of Vinh. While attacking this latter bridge, heavy AA fire erupted, and since they had expended their ordnance, the crews were more than ready to vacate the area and thread their B-57s toward Da Nang for a fuel stop. Navigators Dick Ryan and Tim Cockburn debriefed the mission at Da Nang, indicating probably success, however a more reliable observation was hampered by the heavy AA fire. In the weeks that followed, only six missions of this type were flown in North Vietnam while the B-57s were assigned at Bien Hoa.

Night missions were considered hazardous for various reasons. Vertigo was a primary cause arising from the loss of visual reference with the horizon by the pilot. Other factors included high terrain, inaccurate maps, and reduced visibility due to the bright flares destroying the crew's night vision. Many times the crews that flew these missions had to fly two or three nights consecutively which when coupled with the fact that they had to sleeping 90deg heat during the day produced a very special hazard. It was due to the dedication and skill of the crews involved that no losses were experienced during these initial night operations.

11

The Bien Hoa Holocaust

For the crews of B-57s that flew in Vietnam during this time, the name Bien Hoa has unpleasant overtones that bring to mind a devastating, never to be forgotten disaster. This tragedy occurred on Sunday morning, 16 May 1965 when Capt Charlie Fox and his navigator Vern Hayes were about to start engines and lead Jade Flight, a flight of four aircraft, on an out-of-country strike. The time was 08:15hrs and three other B-57s making up Paget Flight had already taxied out for takeoff. Without warning, Fox's airplane, fully loaded with four 750lb GPs under the wings and nine 500lb GPs in the bomb bay, exploded in a ball of flame with a concussion that threw debris into surrounding aircraft with shattering force. Capt Howard Greene in number four of Jade Flight was away from the rest at the east end of the ramp. He saw the white cloud of smoke and both he and Lt Burbank ducked to avoid fragments, then abandoned their aircraft and ran past the fuel bladders, leaped over the concertina barbed wire (with their seat-parachutes still on) and reached safety in the ditches east of the ramp. The flying bomb fragments and cannon shells caused a chain of explosions that enveloped the ramp in less than one minute. The members of Jade Flight – strapped in their cockpits awaiting engine start time were: Billy Shannon, Jim Underwood, Charlie Fox, Vern Haynes, Art Jepson and Lee Wagner – all perished. Capt Andy Kea was killed by fragments as he was walking from his aborted B-57, having been number four in the flight that just departed. His navigator, Lt Barry Knowles was knocked flat by the blast and received a severe cut across his nose and face. Unable to see, he was rescued and driven away by the maintenance line chief, Senior Master Sergeant Leon E. Adamson who himself suffered multiple shrapnel wounds and severe burns. After a critical and long recovery he was later given a medical discharge. Sergeant Adamson's action illustrates the dedi-

A tightly parked condition prevailed at Bien Hoa even after the mortar attack in November 1964. To have constructed revetments would have necessitated reducing the bomber force at Bien Hoa by half. Unfortunately, that half was destroyed on 16 May 1965 when a ramp explosion took 10 B-57s and killed 28 servicemen – a high price to pay.

Another row with a mix of 8th and 13th Bomb Squadron Canberras stand ready at Bien Hoa for missions against the Viet Cong. Tightly packed conditions seen here were invitations for rocket attacks at night against the Americans. Sufficient ramp space was always a critical issue throughout the war. (USAF)

cation of the maintenance and armament men of the Air Force and the Canberra squadrons that gave untiring service throughout the years.

The explosions hurled fragments as large as coke bottles into the cantonment area almost a mile from the ramp. An entire J65 jet engine was thrown a half a mile. Ten B-57s,* 11 VNAF A-1Hs and one Navy F-8 Crusader which had landed only minutes before the explosion were destroyed. Besides the seven crew members killed, 13 members of the 405th ADVON 1 maintenance squadron perished, and eight Vietnamese from Bien Hoa brought the total dead to 28, with 105 wounded. The base was a disaster, and four B-57s returning from a raid were diverted to Tan Son Nhut as were the three that had taxied prior to the blast and had escaped damage. Casualties may have been even higher for ground crews had it not been for the arrival shortly before the blast, of an unfamiliar Navy plane, the F-8 Crusader, which attracted curious Air Force crews away from the bombers.

*Nos 568, 867, 871, 873, 893, 904, 913, 915, 930 and 937.

12

TAN SON NHUT

To appear not to miss a lick, the men of the Canberra squadrons in Vietnam immediately set up shop at Tan Son Nhut Air Base with the seven aircraft that were diverted there after the Bien Hoa explosion. Ground crews from Bien Hoa were moved to Tan Son Nhut by bus later that day and in four days the bombers were again attacking the enemy. Their use was only on a limited basis, primarily to show the Viet Cong that their 'hated' Canberra squadrons had lost none of their punch.

Realistically, though, the force was badly hurt. The loss of aircraft, crews, maintenance personnel and equipment, all contributed to the reduced capability which plagued ADVON 1 for several days. Their past record was hardly anything to be ashamed of however. The wartime rate for the 18 B-57s maintained in Vietnam was placed at 378 sorties, yet they actually flew 552, which was 41% above the wartime rate. Each B-57 was expected to fly on the average of 84hr a month as a wartime figure, giving ADVON 1 a total of 1512hr to fly for the airplanes assigned. Instead, they flew 2,200hrs, including two courier aircraft, which was 46% above the wartime rate. The men of the 8th and 13th were tigers, and nobody knew it better than the enemy.

Operating out of Tan Son Nhut (TSN) as a combat base was not easy for a number of reasons. This was Saigon's international airport which not only handled vast amounts of commercial traffic, but was also a major military base with an over abundance of varying types of commercial and combat aircraft. One of the most chaotic in the world with its constant flow of arrivals and departures, the airport handled a million pounds of cargo daily. It was two-thirds surrounded by thickly populated suburbs of Saigon, all within easy mortar range of the runways. Fifteen thousand civilian employees meandered around the base daily, making security a night-

A lineman guides 'Red Lima' out of the parking spot and sends it off on another strike against the enemy. The maintenance abort rate for B-57s was amazingly low compared to other aircraft types. This not only spoke well for the airplane, but for the dedication of the men who maintained them. (USAF)

Due to heavy ground losses of B-57s in Vietnam, 20 B-57s were drawn from a variety of duties within the Air Force and returned to Martin in late 1965 to be readied for combat. Among these were 12 B-57Es, originally assigned to tow target duties. Under the tail of the aircraft in the foreground can be seen engine tailpipes, aft nacelle sections, tail cone, elevator and aileron. (Martin)

marish headache. Faced with this congestion, Canberras returning from a strike had difficulty not only in entering the heavy flow of air traffic, but delays on the heavily travelled taxiways were costly in time. Ground crews needed access to the B-57s as quickly as possible for they had to reservice and rearm some of them for a second and often a third mission in the same day. Because the airport was so saturated with airplanes, the B-57s had to be parked on PSP (pierced steel planking) pads. Due to the possibility of ingesting dirt or other foreign objects into the air intake, it was necessary to tow the aircraft on to a taxiway for engine start. This not only made operations slower, but it blocked other traffic as well.

Two tents were pitched by a taxiway to house both operations and maintenance, and the dirt and humid heat made working conditions almost unbearable. Although war is certainly a deadly serious business, humor is also important at times when men, witnessing the fall of their comrades, but never growing accustomed to it, must continue on their assigned missions. Such a saving sense of humor flashed briefly when a high ranking officer approached the B-57 crews with a question a few days after the unit moved into its reduced operations at TSN. 'Do you really mean,' he asked, 'that you are running a combat operation out of these tents on a shoestring like a bunch of gypsies?'

"Yes sir,' he was told, 'but to tell you the truth, we need the work.' Morale was on the upswing.

Improvements with the situation were few if any, but like most difficulties, the crews learned to live with them. The operation was to be a temporary one as the plan was to move the B-57 operation back to Bien Hoa as soon as base repairs were made.

In the meantime, two Air National Guard units at Reno, Nevada, and Louisville, Kentucky, must have been very busy preparing seven and two B-57s respectively for ferry flights across the Pacific as replacement aircraft. In just four days after the Bien Hoa explosion, these airplanes were en route to Clark Air Base in the Philippines. It was at this time that a critical shortage of bomber type B-57s were beginning to be felt within the Air Force inventory. Having an ear always turned to things concerning B-57s, I heard that The Martin Co. was asked to once again resume production of the B-57. They could have used the work – but they de-

A B-57 with its bomb bay door already rotated and open peels off for an attack. On the door are six 500lb GPs fore and aft, and three 260lb frags. On the wings are also 500lb GPs. This unidentified Canberra sent to Vietnam as a replacement was once an ANG RB-57B as determined by the sighting window under the nose. (USAF)

A belly full of bombs, plus loaded wing stations means a lot of trouble for the enemy. Canberras normally dropped one or two bombs per attack pass. With two to four B-57s in the attack pattern on one target, this immobilized the enemy often permanently – during what seemed to them as an endless attack and rain of bombs. Note smoke stains under wing from firing 20mm cannon. (USAF)

Wheels begin to retract as 'Red Hotel' leaves the runway on another mission. The presence of Canberras from these two squadrons were so normal in Vietnam throughout the war that tactical call signs gave way to 'Red Bird' and 'Yellow Bird' for the 13th and 8th BS respectively. (USAF)

The flat, swampy Mekong Delta country spreads out beneath this flight of USAF B-57 Canberras as they return from a strike against the Viet Cong. Weather was tropical, normally with low scattered clouds which often hampered air strikes.

clined, primarily because jigs no longer existed for the airplane that had been out of production for more than 10 years. It would have been too costly to have made new jigs for limited production. The story also got back to me that one general officer in the Pacific suggested using RB-57As in the ground support role to replace losses. This sounded like history repeating itself, having to learn again that the rounded canopy would not optically handle a gunsight. Fortunately nothing more was heard of this suggestion.

These measures on how to prolong the use of B-57s made one point very clear. Aircrews can be trained to fly another type of aircraft for the mission that the B-57 was filling, but it became evident that the efficiency of the B-57 itself had no replacement.

While temporarily at Tan Son Nhut, Canberra crews flew an average daily rate of eight daytime sorties within South Vietnam and three missions per night over North Vietnam on armed reconnaissance. One of these crew members was Maj Frank 'Smash' Chandler, a veteran navigator with the 13th Bomb Squadron. "Smash' tells of an interesting operation when Canberra crews were among the first called in to aid the stricken Special Forces camp in South Vietnam at Dong Xoai. This was in June 1965, when the camp was partially overrun by a regimental size force of the enemy. The VC armed with a staggering array of 81mm mortars, machine guns, recoilless rifles, and flame throwers, had taken advantage of the low clouds and darkness to launch an attack. By morning, when airpower could be used, the enemy had overrun the camp's airstrip and occupied half of the camp itself. In spite of a 900ft scattered-to-broken ceiling obscuring the target, the Canberra crews elected dive bombing to ensure accuracy. The initial drops destroyed six Viet Cong anti-aircraft sites. As a result of this B-57 flak suppression, the A-1 Skyraiders attacking the approach routes to the compound were able to halt the Viet Cong drive temporarily. The Forward Air Controller at one point called for tactical support against 'anything moving outside the compound walls.' This close air support was later credited with breaking the enemy attack and reducing friendly casualties. One B-57 pilot, Capt Russell P. Hunter, was wounded during the Dong Xoai battle when a .50cal bullet entered the bottom of the fuselage, ricocheted several times, finally striking him. In shock, severe pain and bleeding profusely, he 'babied' his aircraft back to Tan Son Nhut, his navigator talking to him continuously to keep him conscious. Following a successful landing he was pulled unconscious from the cockpit and taken to the hospital. What happened after that has become a legend with the action in Vietnam. Russ regained consciousness in the emergency room and slipped away unnoticed He somehow made his way to the flight line and was apprehended in a severe state of shock trying to get into another B-57. All the way back to the hospital he demanded to be released because 'I know where they are now and I'm going to get them.' (This courageous pilot was shot down in Laos seven months later and not recovered.)

The B-57 was proving its ability to sustain combat damage and return its crews to safety, although there were some that did not make it back. It was a noteworthy occasion when the first crew ejected from a B-57 after sustaining major damage and was safely recovered. This first incident occurred on 8 June 1965 when 882, piloted by Capt Gordon Nelson with his navigator Capt James R. Carnes, was hit on a strike mission in the Delta west of Can Tho. Here is Nelson's report of the incident:

> 'I was engaged in a close air support mission, about 80 miles south of Saigon. I was number four in a flight of B-57s. As I rolled in on my third dive bomb pass, the left wing burst into flame. After advising the flight that I had been hit, I shut down the engine and activated the fire extinguisher. The fire did not go out. Capt Carnes then said he thought the fire was in the wing tank. We decided we had done as much as we could, and that it was time to leave the aircraft. Carnes was to blow the canopy and eject first and I would go as soon as I could after the canopy went. Jim delayed slightly and I went before him (he later said he was marvelling at how well the canopy came off). My equipment worked perfectly, but my lap belt did not disconnect. I found this out later as the horizon was spinning by my eyes and the seat would not kick away. I pulled the belt manually and then the "D" ring. The canopy opened and deployed perfectly, and I removed my mask and looked down at the craters from our bombs. I was expecting the Viet Cong to shoot when suddenly two armed Army helicopters were on each side of me spraying the woods with machine gun fire. I slipped the chute to get on the other side of a small canal and landed on a grass house in a village. The village was deserted and after getting out of my gear I saw a smoke grenade that had come from one of the choppers. It landed in a clearing near the village and I ran toward it. Jim was already on board. It took four minutes and nineteen seconds to get us – it seemed like four hours!'

According to John Kendrick, who was the lead navigator for the flight led by Ben Stone, they spotted the .50cal gun position at the side of the attack course that had caused all the trouble for Nelson and Carnes. John wished it to be recorded that 'due punishment was dealt to the gun position and its occupants for downing their friends and any that might follow.'

All losses are tragic, some more than others because of the circumstances and kinship felt with the men involved. One of these occurred on 29 June 1965, following an air strike in support of Vietnamese Rangers in contact with the Viet Cong. While on his last strafing run, 'Bud' (Samuel P.) Chambers in 895 was hit by ground fire, but there were no serious indications of trouble. En route home, his wingman observed several six-inch holes in the left wing and a few close to the compressor section of the engine. Fuel was streaming from the wing although Bud reported that all indications in the cockpit were normal. A straight in approach was established for Tan Son Nhut. The gear was lowered and everything was still normal as verified by the wing man, except that fuel continued to come from the left wing. A safe landing seemed assured as Bud was on final approach and he was cleared to land. As the wing flaps started down, the flow of air was disturbed and the streaming fuel was sucked through holes near the hot engine and exploded, causing the left wing to break off the airplane. Bud and Bob Landringham, his navigator, ejected instantly at a low 700ft altitude – as the airplane rolled inverted. Bob struck the ground while still in his seat, and Bud's chute was seen to stream out as he hit the river. His body was recovered two days later. This incident touched so many, that a new BOQ building being constructed at Clark AB was named 'Chambers Hall' in his memory.

13

Da Nang and the DOOM Pussy

The move back to Bien Hoa from the six weeks' stay at Tan Son Nhut did not materialize. Instead, the B-57 operation was moved in late June 1965 to Da Nang Air Base, a location much closer to the DMZ. This was a growing base, the largest in the northern portion of South Vietnam, situated on sandy terrain just outside the big port city of Da Nang. This new location provided greater working room for the Canberras and better living conditions for their crews. For the 8th and the 13th, this was the beginning of a much smoother running combat operation. When deployed at Da Nang, they came under the operational control of the 6252nd Tactical Fighter Wing, which became the 35th TFW about a year later.

The most noticeable change in the operation was the placing of one entire squadron of B-57s with all their support at Da Nang

Da Nang was a growing base, but with the move there came a stabilizing effect on the B-57 operation. This view taken in late 1965 looks west with revetments in the center to house the Canberras.

The move to Da Nang in June 1965 took place before aircraft revetments were constructed. Ramp space was more in keeping with bomber operations than on earlier bases. Wing racks of these B-57s hold 750lb tritonal filled bombs. (USAF)

on TDY (temporary duty) away from Clark for 60 days at a time, while the other squadron remained at Clark and accomplished training for their return to Da Nang. This provided a more equitable arrangement for the crews and continuity in the combat zone, for there were no daily crew rotations to contend with.

This system began on 21 June 1965 when members of the 13th left Tan Son Nhut and were replaced by a contingent of the 8th Bomb Squadron. They were to join the parent organization at Da Nang when it moved there from Clark as a unit. That date slipped from 28 June to sometime in July due to delays of a F-100 unit being transferred to another base. In anticipation of this move and to reduce the distance to and from targets in North Vietnam, the night bomber force of the 8th had moved to Da Nang on 18 June. Col Danny Farr was in command at this time, having taken that position the previous month. This rotation of complete squadrons brought an end to the acronym ADVON 1, which symbolized only a composite of both the 8th and the 13th Bomb Squadrons. The B-57s were so well known on the tactical scene, that after a few months, nondescript call signs such as Sofa, Kay, Newt, Dark, etc, gave way to the more affectionate sounding 'Yellow Bird' or 'Red Bird' for the 8th and 13th respectively, depending on the squadron color of the unit then in-country.

Missions of varying types for the B-57s made the war anything but a boring pastime. One of these was flak-suppression missions with the C-123 Ranch Hand aircraft. Americans, by nature, are not jungle fighters. To enhance our advantage to fight the war on our terms, the jungle growth along the enemy's supply routes was sprayed with defoliate, thus depriving him of this protective visual cover. These defoliation sorties were flown low and slow over some of the roughest terrain in the world for this type of flying. Hits from ground fire were the rule rather than the exception.

The presence of the Canberra in company with the Ranch Hand aircraft was a deterrent factor by causing the enemy to think twice before opening fire. When ground fire was encountered, a crew member in the C-123 would toss out a smoke grenade marking the spot and the B-57s would unmercifully saturate that area of the jungle with their guns and 260lb fragmentation bombs.

The enemy's special hatred for the B-57s was often evident in the testimony of VC defectors. When the VC were interrogated by American-South Vietnamese intelligence as to why they defected, many replied that the primary reason was overwhelming American and Vietnamese air power. They exclaimed that there was no place to hide, and were always kept on the move. Their tactics were continually being modified because of this, and the pounding from the air was unbearable. When interrogators showed defectors the silhouettes of aircraft, most of them immediately pointed to the B-57 as the most despised. 'The screaming bird,' they would say. 'It is the worst. It stays over the target so long. And it never runs out of bombs.'

Weapons so dreaded have a psychological assist – for example, the Stuka dive bomber of World War II, with its deadly accuracy and awesome scream. The J65 engine of the B-57 also has a weird whine all its own. The VC had a contemptible word for the Canberra *can sau*, which meant 'caterpillar', the little creature that they considered the most loathsome thing in all Vietnam.

Bombs carried by the B-57s throughout the war were fairly consistent in type, but load configuration often varied on the mission and bomb type availability. The wing stations normally carried 4 x 750lb GPs (General Purpose) or 4 x 750lb napalm tanks. On occasion, 500lb GPs, 260lb fragmentation bombs, rocket pods for 2.75in 'Mighty Mouse' rockets, 5in rockets, and night flare pods were carried on the wing pylons. Internal loads ranged from 4 x 750lb GPs, 9-13 x 500lb GPs, or 16-28 x 260lb frags. Mk 81 and Mk 82 bombs with either high drag or low drag fins were common. For some special missions, 1,000lb HEs (High Explosive) were carried. On out-of-country missions, M35 and M36 'Funny Bombs' were most memorable to all the crews that carried them. These fire bombs spread a blanket of hot burning thermite over a large area that sustained their fire for up to 10-15 minutes, lethal for trucks and area targets. From February to April 1966, B-57s were consistently sent out with only one light bomb and half the guns charged. This retained the show of force while the heavy bombs were diverted to fill shortages created by the start of the B-52 raids.

Mission flying times also had a wide range variation. Attacks against designated targets in South Vietnam could be under two hours but averaged two and a half. To the more distant out-of-country targets the average was three hours or more. Night missions into North Vietnam and along the Ho Chi Minh Trail became more numerous while the Canberras operated from Da Nang. As time went on they also became the most hazardous to fly because the abundant radar controlled anti-aircraft guns operated more freely under the cover of darkness. The B-57s were so effective against enemy supply routes at night that it seemed to justify the risk.

Only the more experienced crews flew these missions due to their many complexities. It was not long before these night sorties *up-north* were even given a special and ominous sounding name –

This unusual picture taken in early 1966 shows a Douglas EB-66 leading a formation of B-57s deep into North Vietnam at the instant of bomb release. This was a 'Combat Sky Spot' mission where radar was used for target acquisition and command for bomb release is given by voice or electronic signal. EB-66s could detect active missile sites for the formation. Canberra missions of this type, however, were few and were replaced by B-52 raids. (D. Beggerly)

The Doom Pussy patch was worn only by those more experienced crews who flew night missions 'up north.'

These trailers provided air conditioned and less noisy sleeping quarters for the four crews required to fly each night. At Da Nang even crew facilities began to improve. (D. Beggerly)

'Doom Pussy Missions.' Crews that flew these missions were entitled to wear the 'Doom Pussy' emblem sewn on the right shoulder of their flying suit. This was an embroidered head of a wide jawed yellow cat with pointed ears and a black patch over his right eye. Clenched in his teeth was the outline of a twin engine aircraft – very much like that of a Canberra. In green letters around the border were the words: *Trong Mieng Cua Con Meo Cua Tham Phan*, which translated from Vietnamese, reads: 'I have flown into the jaws of the cat of death.' Returning crews, having had a night of it up-north, dodging red-hot 37 and 57mm tracers and technicolor air bursts that seemed to envelop the airplane, simply phrased it: 'I have *seen* the Doom Pussy.'

This legendary cat got its name from the Da Nang Officers Open Mess, thus: DOOM. On the back of the bar was a tall and slender wood carved cat, placed there for no apparent reason even before the B-57s arrived at Da Nang. While fellow crewmen were up-north on night missions, evading Russian made Surface-to-Air Missiles (SAM) and radar aimed guns, the cat was turned to face the wall until the safe return of that night's Doom Pussy Missions.

14

A Few of the Rough Ones

North Vietnam at night was bad enough, but being there during daylight hours was something else. The situation takes on a different perspective however, when there is the chance to help a downed fellow American aircrewman. This was the case when B-57s were called upon for the first time to provide RESCAP (rescue combat air patrol) for a downed F-105 pilot in North Vietnam on 27 July 1965. His downed position was 29 miles west of Hanoi – right where nobody wanted to be, due to the presence of SAMs and automatic AA in this heavily defended area.

Lt Col Danny Farr, Commander of the 8th, with Billy Boyington as navigator were off the runway at Da Nang at 16:50hrs in response to this emergency. On their wing was Bob 'Whale' Smith with John Hughes in the back seat. The flight of two were vectored by Panama, the radar ground station, to the downed pilot's position and arrived there an hour and eight minutes later after being vectored around known flak positions. The plan was to stay at altitude to conserve fuel, for assistance would not be needed until the arrival of the rescue helicopter. They would then provide fire suppression during the pickup phase.

Time over this hot area dragged on and still no chopper. Col Farr and his flight conserved fuel as best they could but they finally had to leave the area and recover at Udorn, Thailand, which was the nearest airfield at which they could refuel and be ready to assist further if need be. Darkness was fast approaching, but fortunately a

Pulling up from this attack, Yellow Bird 31 of the 8th Bomb Squadron just released its bombs on a Viet Cong concentration in the area of the smoke at upper left. This was the critical point for the Canberra when closest to the ground and subject to ground fire. (R. Hunter)

second flight of B-57s had taken off 20 minutes after the first as a backup. Gail Manning was leading with Ken Blackwell on his wing. This flight with Phil Mason and John Kendrick as navigators was vectored to intercept the helicopter as it crossed the Laotian border and flew escort with it to the downed pilot's position. It was a race against time now, for shadows were getting longer as daylight was coming to an end. Fortunately the F-105 pilot's radio bleeper was sending a strong signal for the rescue aircraft to home in on and little difficulty was had in locating him. The chopper was being fired on but the crew was too busy attempting to unjam the rescue hoist to direct the B-57s to the point of fire. The hoist refused to cooperate so they had to land in a nearby clearing so that the pilot could climb aboard.

During this time the two B-57s became the center of attraction of the local AA guns and received some damage which perhaps drew the enemy's fire away from the rescue operations. The rescue was completed with no time to spare, for as soon as they were able, Gail Manning took his flight direct to Udorn, where they landed with only emergency fuel remaining.

Missions up north in the relatively slow flying Canberra were not one of the best occupations for these aircraft in this heavily defended area. The only satisfaction with this type of mission was in knowing they had aided in the safe recovery of one of their own people. Perhaps this was the incentive for the Jolly Green rescue helicopter crews doing this on a daily basis; a job considered the most hazardous of all flying jobs in combat.

Then there was the case where one B-57 made a very deliberate attempt to shoot another B-57 out of the sky This unusual situation took place on 6 August 1965 when 919 received extensive battle damage during a daylight strike near the seacoast town of Nha Trang along the central coast of South Vietnam. It was all that Larry Horacek could do to keep his crippled bomber under control before he and his navigator Bob Johnson would have to eject. Due to the population in the area, Horacek wanted to have his B-57 headed toward the open sea before leaving it, especially since it was still loaded with bombs that he could not release. As they crossed the shoreline, both crewmembers ejected, each making a safe water landing and were picked up within a few minutes by a helicopter.

This is not the happy ending of the story, however. The stricken Canberra had ideas of its own. Apparently the left engine lost power, for the airplane began a gentle left turn which if continued would turn it back towards land. In desperation, the pilot of the second B-57 in the flight, closed on the runaway aircraft and began firing, hoping to shoot it down while still over water. Guns on Canberras are not aligned for air-to-air combat – but are deflected downward slightly for ground strafing. This attempt to achieve an air-to-air kill was unsuccessful and the pilot had to break off the attack for fear his rounds would strike inland as the crippled bomber had now

Clark Air Base in the Philippines was safe haven for the 8th and 13th Bomb Squadrons which deployed from here on temporary duty to the combat zone throughout most of the war. The base provided rear echelon maintenance for the Canberras on a time cycle for the rotational squadron at Clark. This was the B-57 ramp supported by their parent organization, the 405th TFW. Squadron Operations building is center rear. (D. Beggerly)

The results of a hot night over Tchepone, Laos in March 1966, and proof that the structure of the Canberra could take punishment. This B-57 was ready for combat the next day, having had its stabilizer replaced by one from 550 which bellied in from combat damage a few nights before. (E. Youngs)

Another view of 905 being checked over after a safe return to Da Nang. Fortunately for Art Kono and Paul Stenseth, they were far enough ahead of the antiaircraft round that otherwise could have hit the engine with more disastrous effects. (USAF)

completed its turn. Moments later the pilotless B-57 slammed into the heart of Nha Trang, killing 14 Vietnamese, injuring 67 others, and also injured eight Americans. Four of the bombs exploded on impact, but fortunately the other 12 did not.

Eight months later, Larry Horacek was shot down again, this time while escorting a defoliation mission north-west of Dong Hoi, in North Vietnam. As before, he was rescued by helicopter, and finished out the remaining months of his tour – *very carefully*.

Combat missions, like anything else repeated frequently, can be expected to become routine by following a fairly consistent pattern. But when standing "Flaming Arrow"* alert and ordered into the air, anything could happen. And happen it did when such an alert was sounded, asking for close air support to the beleaguered Plei Mei Special Forces Camp situated in II Corp.

This mission began in the early morning hours of 20 October 1965, when 8th Bomb Squadron crews Captain Howard A. "Shakey" Steady, with navigator Lt Mike Chaloult, along with Lt Art Kono and navigator Lt Paul Stenseth responded to the launch order. Turning south out of Da Nang, Plei Mei was about 20 minutes away. They were the first aircraft to reach the scene where the enemy was trying to overrun the camp. Their air strikes were to be hampered by a low, 3000-4000 foot ceiling. Above the clouds was a C-47 dropping flares over the camp and brightly illuminating the clouds. With the enemy at the camp perimeter wire between the camp wall and the runway, the FAC requested the two B-57 crews to drop their bombs along and parallel to the wire, but NOT to hit the dirt runway or damage the camp barrier wall. This made for a very narrow confining strip.

*"Flaming Arrow" was air support for American forces only. Before radio contact was possible from remote special forces camps, a large arrow inscribed on the ground was set on fire to point in the direction of attacking forces.

Very little publicity was given to the B-57s and their crews because so many of their combat missions were of a classified nature in Laos, Cambodia and North Vietnam. Consequently it was a little known fact that these Canberras operated continually in the air war for more than five and a half uninterrupted years, then for two more years as B-57Gs. (USAF)

This is one of 20 B-57s reconditioned in 1965 by Martin to replace combat losses in the war zone. From test projects to combat, 499 is shown here in war paint and has 750lb napalm tanks on the wings. Canberras took on camouflage beginning in December 1965. Note openings for four .50cal machine guns outboard of the engines. This Canberra, 52-1499, is now part of the Air Force Museum collection. (USAF)

Unable to glide bomb in the B-57s usual weapon delivery mode because of the low ceiling, level bombing had to be done by the pilot toggling off the bombs by his own calculations based upon experience. Each Canberra carried on their bomb door 21 260lb frags each. After both airplanes made two passes, dropping all these bombs while under a white silhouetting ceiling which made them easy targets for the enemy, the ground attack had been stopped cold. Fortunately, the enemy had not fully established their .50cal. AA gun position on a rise south of the camp, allowing the Canberras to strafe the retreating insurgents that were also well lighted from above. The crews were then directed by the FAC to drop the remaining 4 500lb G.P. bombs that each B-57 had under its wings, on an enemy suspected staging area.

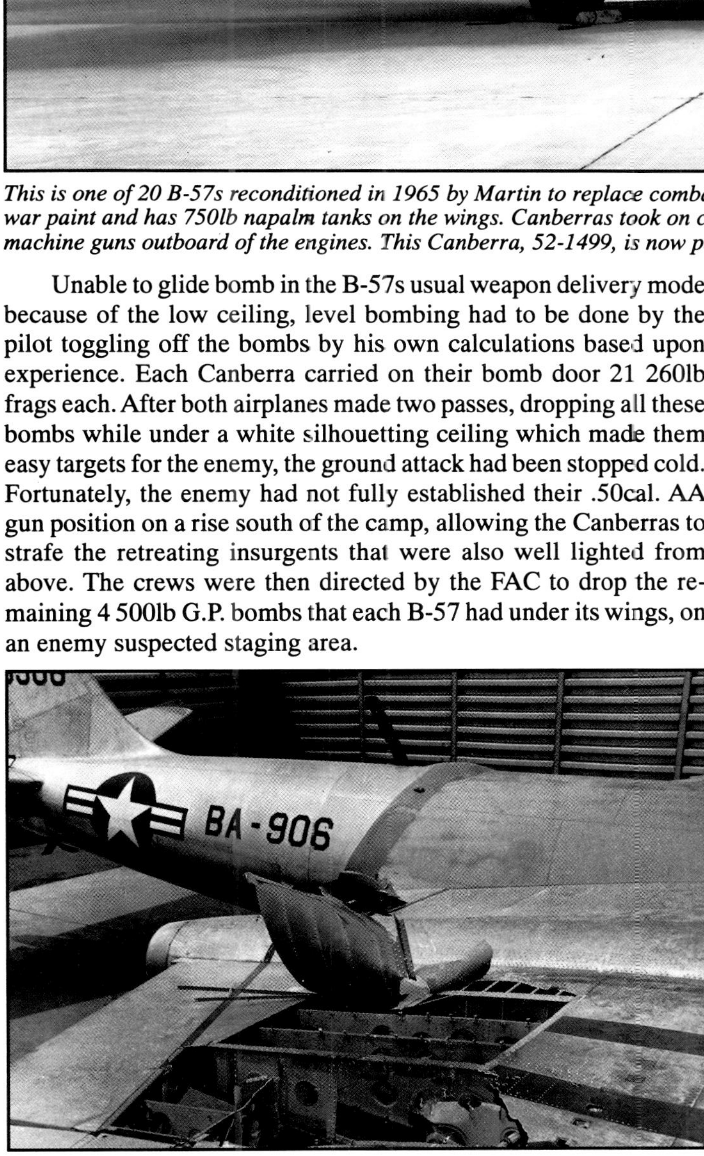

The Canberra could lose a lot of wing and still continue to fly as proven in this view of 906. It was hit in several places throughout its structure over Tchepone, Laos, on 15 March 1966. Cabin pressurization, electrical and hydraulic systems were also knocked out, yet 906 was able to return to Da Nang. (L. Mongeon)

The Plei Me fight appeared to be over, but this was not the case. Soon after day break, Tout 21 and 22, two other 8th Bomb Squadron B-57s left Da Nang on what was a routinely scheduled in-country strike on a preselected target. This was to be in I Corps along the 'Street Without Joy', north-west of Da Nang. Soon after becoming airborne, the Canberra crews routinely changed their radios to the Direct Air Support Center's frequency and checked in with Big Control. This agency had operational strike control over tactical aircraft in the I Corps area, and would pass along any new information pertinent to the strike mission.

The routine of their planned mission ended here however, for as soon as contact was made with the DASC, they too were given a 'Flaming Arrow' divert to Plei Mei. A FAC was on the scene, ready to direct attacks. Things had happened since the first B-57s had left, for smoke from a downed helicopter and U-6 Beaver readily identified the spot for the two B-57 pilots. The Viet Cong forces were now estimated to be of regimental strength, since they reappeared after the first B-57 strikes. They did not retreat under the helicopter air attack, and had already inflicted heavy casualties on the camp. Heavy ground fire was observed by both Canberra crews and was also reported by the FAC. It was later learned that this was the first major battle with North Vietnamese regulars that would now stand and fight.

Cloud cover in the area remained at 4,000ft so these crews too had to improvise with level bombing runs at 2,200ft and 350kts. They released one wing bomb at a time on known enemy troop emplacements. After four or five passes by each aircraft, they initiated strafing runs covering strategic areas of concealment where the enemy was likely to have vantage points.

In the lead aircraft, 920, was Maj Jerry Hamilton with Maj Harold Holzapple as his navigator. On their third strafing pass, the Canberra took numerous hits from the south ridge which now had .50cal machine guns in place and responding. A bright orange ball of fire appeared briefly over the right wing and the tip tank and part

Pilot Larry Mason assesses the damage to the right wing of his B-57 after recovering at Da Nang. This severity of damage is evidence that the structure of the Canberra was designed to take the ultimate in punishment and be able to remain in the air without further failure.

of the right wing left the airplane. The B-57 was seemingly still under control – for the time being at least. Maj Doug Beggerly and Lt Dick Lewis in the second B-57, 931, having seen where the ground fire came from, directed their next strafing pass on that position. As they pulled up from the run, they took a hit with a resounding explosion in the cockpit between the two crewmember seats. The cockpit immediately filled with smoke. The hit rendered the oxygen system useless, making breathing under these conditions almost impossible. When Beggerly got the cockpit cleared so he could see and breath freely, it became apparent that his electrical navigation equipment was also inoperative.

But now too, Jerry Hamilton's problems were beginning to multiply. With a jaggedly split right wing tank and a hung bomb on the left wing station, the airplane became uncontrollable when the airspeed was allowed to decrease. Nothing would jettison, not even the full internal load of 260lb fragmentation bombs. There was no other recourse than to bail out while the airplane was still under control. Setting the airplane in a climb as they headed it toward Pleiku, where rescue helicopters were already summoned, both crewmen punched-out just before the airplane went out of control. Despite the damage to his own ship, Beggerly circled the area of his downed teammates while directing the rescue operation and providing air cover until they were safely aboard the rescue helicopters.

With half the problem solved, Doug could now turn attention to his own. Not knowing the full extent of his damage, he decided to land at Pleiku, which was all but in sight of where this incident took place. His hydraulic system was also at zero which required him to use the emergency hand pump system to lower the landing gear. After a successful landing, there was a welcome reunion for the four at Pleiku, although Maj Holzapple received a fractured vertebra and was hospitalized at Pleiku, and later air evacuated to the States. Egos were a bit ruffled when the three members of Tout Flight returned to Da Nang without their Canberras, while seated in the back end of a C-7 Caribou.

In a letter that Doug wrote to me the same day of the incident, he started out with: 'this is *now* a personal war! I had planned to keep it cool while I was here, but now they have made me *mad*!' You see – this was Doug's very first combat mission in Vietnam – but it was his third war in the air. For his bravery on this mission, Doug not only received the first Thirteenth Air Force 'Well Done' award to be presented, but also the Distinguished Flying Cross, first Oak Leaf Cluster, for his heroism, and Order of the Able Aeronaut. Crewmembers of all four participating B-57s received the Distinguished Flying Cross – and well deserved.

As for the outcome of the besieged outpost of Plei Mei, six more B-57s entered the attack that morning followed by other strike forces. After two days of intense fighting and helicopter airlift of support elements, the camp was finally secured.

There have been many acts of heroism among crewmembers of the B-57s that were created by combat situations. Not only do the favorable outcomes of these situations reflect well upon the crewmember's knowledge of their airplane and related systems, but also that the B-57 could withstand the punishment inflicted by combat damage.

Another such situation occurred to 906 of the 8th Bomb Squadron, on 15 March 1966, near Tchepon, Laos. Capt Larry Mason was on a strafing run on enemy trucks when his Canberra was hit by anti-aircraft fire. The damage was so severe that the aircraft rolled almost inverted but held together. After regaining control of his aircraft, Larry's first thought was that he and his navigator Capt Jere Joyner, would have to eject. His cockpit indications showed loss of power on one engine and a fire warning light on the other. Struggling as he reached forward, Jere passed him a blood-stained message which read: 'Hit badly – arm and leg – losing blood.'

Realizing that Jere possibly would not survive bailing out, Larry passed him a tourniquet and gingerly headed his crippled and radioless B-57 to Da Nang. He was successful in reaching the base, but the landing gear indicators showed the left main and nose gear in the intermediate position and the right main gear down. Unknown to Larry was that one of the shell hits caused all three gears to drop down and lock, while the cockpit indication was erroneous. Pressed with getting his navigator to medical aid, yet unable to get a safe gear down indication, Larry placed the gear handle in the up position on this third pass at the field and made what he thought would be a gear-up landing. To his amazement, the aircraft landed smoothly on the extended gear and made a normal rollout. For this heroic outcome that saved the life of his navigator, Capt Mason received the Thirteenth Air Force 'Well Done' Award, the USAF 'Well Done' Award, the Koren Kolligian Jr Trophy for 1966, the Order of the Able Aeronaut, and more importantly – the Air Force Cross, the only AFC connected with B-57 operations.

A postscript to this harrowing story is that the Canberra also survived this encounter, thanks to the crew, and the ground maintenance personnel that healed its wounds. After nearly three more years of combat, it was modified as a B-57G and was again returned to combat.

15

VNAF FLIES B-57S

Front page headlines on the Saigon Sunday Post for 1 August 1965 proclaimed: 'B-57 Bombers For VNAF.' This first public announcement stated that 'The Vietnamese Air Force would soon have its first jet bombers,' and that 'four B-57 twin jet Canberra bombers will soon be turned over to them.' The United States was reluctant to equip VNAF (Vietnamese Air Force) with jets for fear of further escalating the war. However, the U.S. was reminded that they had equipped other nations friendly to them with jets, so this pressure from Saigon, coupled with a need to boost morale of the South Vietnamese people prompted new thinking. The consoling factor at that time for U.S. war planners was that the jets in VNAF would be a token force at the most. To what extent

Reflected pride is obvious, especially for VNAF Captain Lan, having just landed at Clark from his first solo ordnance delivery training mission at Crow Valley Range in the Philippines. John Kendrick, in background, was Stan Eval (evaluation) Navigator and went along, while John Alder, right, had been Lan's instructor pilot. Soon after, Lan was flying combat missions from Da Nang with the Americans. (J. Kendrick)

American Canberras were often re-marked with VNAF colors on an 'as needed' basis. This required the removal of USAF nose and wing markings, PACAF insignia on the tail, adding a Vietnamese flag to the rudder, and modifying the national insignia. VNAF insignia was outlined in red with yellow bars. USAF serial numbers remained as did the yellow fuselage band of the 8th Bomb Squadron. (D. Beggerly)

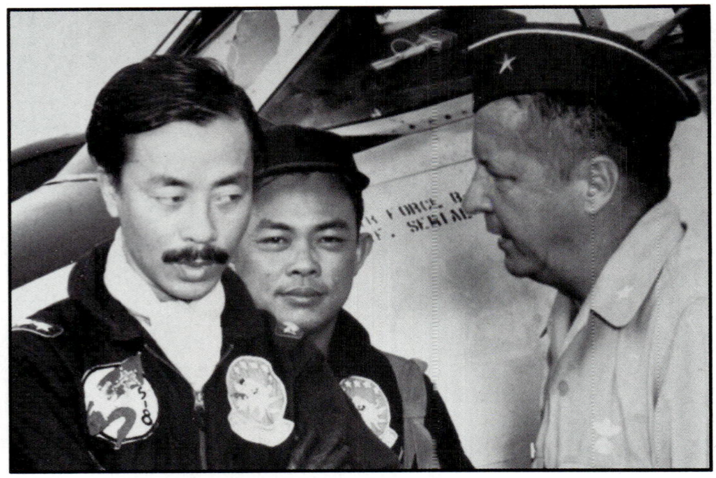

Nguyen Cao Ky, left, Premier of the Republic of Vietnam and Commander of VNAF talks with Brig Gen Albert W. Schinz, MACV Advisor. The occasion at Tan Son Nhut was the presentation of the first jet aircraft, four B-57s, to the Vietnamese Air Force on 9 August 1965. Capt Tuong, pilot of another Canberra looks on. (USAF)

this program was expected to develop must be left to conjecture, but what was not planned was that in less than seven months after this announcement, the VNAF B-57 programm would end, abruptly, and without fanfare.

The program actually got underway as early as May 1964, only one month after the B-57s arrived at Clark. 2nd Air Division, acting as liaison, conveyed the message that called for introduction training in B-57s for six VNAF officers. It was not long before VNAF Maj N. N. Bien, and Capts Tuong and Long arrived at Clark to begin this training. The American instructors were relieved to learn that the Vietnamese pilots had some jet experience in T-33s, either obtained in France, the U.S., or both. Heading this training program was Capt Don Nation, who recalls that after the initial problem of the pilots transitioning from the Douglas A-1 Skyraider to a twin engine jet, he was most emphatic that these officers were excellent pilots. All three wore 500 *combat mission* patches which gave no indication as to how much additional experience lay behind them. (The counting of combat missions ended after 2,000 unless they received battle damage.)

Training for the next three officers proved to be very interesting also. Unlike the first group of three, this training – which the directive message referred to as 'Familiarization Training' – was to be accomplished at Tan Son Nhut instead of Clark. It soon became obvious why, for these students were none other than Nguyen Cao Ky, Commanding General of VNAF (later to be Premier of the Republic of Vietnam), Col Luan (later Gen), in charge of Internal Security for Ky, and Maj Van, Chief of the VNAF Command Post. With one dual control B-57C and training aids in hand, Don Nation got the program started, thinking 'Familiarization Training' meant about two rides for each of the officers. What 2nd Air Division intended was about 30 hours for each and the program to be accomplished in about 30 days. It was not long before Don got additional help.

The often present low clouds hang on nearby mountains at Da Nang on 29 October 1965, as 541 has just been prepared for a strike against a suspected VC stronghold in central Vietnam. VNAF crews and their aircraft were under the operational control of the American forces and shared the base facilities of the American bomber squadrons.

It was not unusual to see a VNAF marked B-57 on the ramp at Tan Son Nhut. Premier Ky wanted the Canberras with VNAF markings as visible as possible to show the bombing presence of his Air Force. Shown here is 53-3879 along side RB-57E Patricia Lynn aircraft, 29 October 1965.

Ground school was finished in about three days. One and a half days were spent on the very hot ramp with aircraft and cockpit familiarization, and then into the air to apply the knowledge learned on the ground. Flying went on for three or four days and then came time for a change of pace. As guests of Ky, the entire contingent went to the Central Highlands for a bit of Bengal tiger hunting. Don has many yarns to tell of this unusual hunt – on the backs of elephants at night with battery operated spotlights – but that is another story. Tigers were not encountered, but they did manage to bag three deer.

More than a year passed before the announcement was made that VNAF would receive B-57s, and a formal ceremony was arranged for the 9 August presentation to the Vietnamese Government. When those who attended arrived at Tan Son Nhut Air Base, there was already one B-57B, 530, in the spectators area which had been flown down from Da Nang the day before by Majs Sims and McCord of the 8th Bomb Squadron. This Canberra was void of all US markings, except for the shiny telltale area where the letters of 'U.S. AIR FORCE' had once been painted. The Vietnamese insignia replaced the US national marking, and the yellow and red Vietnamese flag was painted across the rudder. In front of the airplane was a display of various ordnance loads that the B-57 could carry. (It was not discovered until the evening preceding the ceremony that one 500lb bomb was already fused!).

Before this ceremonial day, Capts Don Nation, Frank Hardee and Harry Zahn, brought three dual control 'C' models from Clark so that Ky, now Premier and retaining command of VNAF, and VNAF Captains Toung and Long could refamiliarize themselves with the airplanes. On the eve of the transfer day, these three airplanes, 849, 883 and 940 also got the treatment of new Vietnamese markings. The reviewing stands that typical hot, humid, August day were filled with representatives of SEATO nations, ambassadors, generals, admirals and other officials – of all sizes and shapes. The three B-57s had taken off well before the spectators gathered, and at the precise moment specified in the program, they thundered across the field in a high speed, low altitude pass in front of the reviewing stand. In the lead aircraft, 849, about 30 seconds in front of the other two was Gen Ky, with Don Nation in the back seat as were the two American pilots in the other two ships. Ky was an excellent pilot, but since he was unable to participate in the practice fly-bys preceding this event, Don took over at this point for the landing. His humorous comments about the event are worth repeating here:

> 'As we neared the far end of the runway, I pulled up in a steep climb with a half-roll and reverse for what I thought would make an interesting entry onto the downwind for landing. We still had a good head of speed, but I fudged a bit and got the gear down at not too excessive an airspeed at about the time I turned on base leg. While doing everything possible to slow down to flap extension speed, yet keep the pattern in tight, I started talking to myself over the hot-mike for the benefit of the General in the front seat, hoping this would reassure him (and myself) that everything was still under control! "Slow down – you (s.o.b.) – s-l-o-w d-o-w-n!", I expounded, not knowing that my death grip on the throttles, attempting to get the

engines to idle even slower, was also keying the mike button, and my comments were being transmitted. Moments later as I rolled out on final, a very clam American voice from the tower responded to my comments, "Dragon One, if you have your aircraft slowed down sufficiently, you are cleared to land." This transmission which identified my aircraft, made it clear that I had boo-booed! Needless to say, Frank and Harry spent much time pointing out my lack of radio discipline for weeks that followed.'

All three B-57s got on the runway in spectacular form to the delight of the audience (and aircrews!). At the far end of the runway, out of sight of spectators, the canopies raised and the three American pilots slid out of the back seats and on to the ground. The Vietnamese pilots taxied the ships to the reviewing area, knowing that their solo appearance would have a prestigious effect upon the press – which it did.

Ky, looking very dapper as always in his distinctive black tailored flying suit, climbed from the cockpit amidst reporters and flashing cameras. He had insisted that the three American pilots, despite their baggy, sweat soaked flying suits, join him at the champagne reception that followed the speech making. This stunned the high ranking attendants from 2nd Air Division, thinking it would compromise the delivery impressions that were made, but they acquiesced. While festivities continued into the evening, three other B-57Bs at Tan Son Nhut were painted with Vietnamese markings and exchanged for the 'Cs'. These dual control models were again

VNAF marked B-57, 929, taxies past the watchful gaze of an Air Policeman at Da Nang AB. The missing 'U.S. Air Force' normally on the nose is the first obvious indication that this is a Vietnamese Air Force assigned Canberra. One crewmember on board was usually an American, until the Vietnamese gained considerable experience in the aircraft. (USAF)

painted with USAF markings and returned to Clark where they again took up their vital role of training more pilots and navigators.

Shortly after the presentation day, a transition and weapons delivery school for Vietnamese pilots was started at Clark, more openly than the first. Six additional pilots were selected which soon multiplied to names of 12, along with 12 navigators, and a number of very eager VNAF maintenance personnel. Far more interesting

These VNAF B-57s from Da Nang have just rejoined after attacking suspected VC strongholds in central Vietnam on 29 October 1965. American crews were often on board as in this case, but many VNAF pilots and navigators became very effective with their Canberras. For some, however, the airplane was too heavy for them to handle adequately. (D. Beggerly)

VNAF usage of B-57s was at a transitional time when the Canberras were being camouflaged. The colorful VNAF insignia and flag compliment this new scheme very well. This is 52-1532 photographed at Da Nang August 1966. (D. Menard)

is the fact that two Philippine Air Force navigators were included in this B-57 program. The reason was strictly political, for it was a U.S. State Department sponsored program, and in order for the Philippine Government to allow VNAF personnel into their country, they traded the training of two navigators.

The new program began on 20 September 1965 when Maj Bien, who returned for a refresher course and weapon delivery phase, and Capt Lan began flying at Clark. Three weeks later, Lts Ty and Vu joined them. Each pilot was to receive 70 hours in the airplane with no less than 40 training sorties before being considered qualified in the Canberra. With the help of Hardee, Zahn and Alder, training moved into the Combat Weapons Delivery Phase. Nation reflects:

> 'Because of their many years of combat experience already behind them, I think they ended up teaching us in this final phase. They were very good with ordnance delivery. As I recall, it was Capt Long that literally drove Harry "Z" up the wall with his dive bombing. His first bomb was invariably 250-350ft short, then the next ones would be right down the pipe! Long's explanation for this was to scare the enemy into the target area, then really zap 'em.'

Navigator training began 11 October and this responsibility was given to Capt John Kendrick. This training was to familiarize them with the systems they would operate in the back seat of the B-57, and to generally transition their thinking from reciprocating engine aircraft speeds to that of jets. As the crews completed their training, they went to Da Nang and flew their combat missions with either the 8th or 13th, which ever was on station at the time. To gain combat experience in the B-57, each new crewmember flew with an American pilot or navigator, which ever the case may be. Eventually the VNAF crew members were teamed together and flew in VNAF marked B-57s. All the while, their combat missions remained under USAF operational control. There were usually a few B-57s in VNAF markings at Da Nang for these crews, but these ships were changed frequently to meet maintenance schedules for down times.

In the meantime, the VNAF B-57 program was not receiving the visibility that the Vietnamese Government felt it should. The enemy had not felt the impact of this VNAF bomber force, and such action would increase morale of the South Vietnamese. An appropriate time to rectify this situation came on the Vietnamese Armed Forces Day which was on Friday, 29 October 1965. To highlight the occasion, five B-57s from the 8th Bomb Squadron, then on rotation at Da Nang, were repainted with VNAF insignia, and each carried 13 500lb GP bombs and fully loaded guns. En route to Saigon, these Canberras, manned totally by American crews,* conducted a preplanned air strike on a suspected VC stronghold, making the presence of VNAF bombers known before landing at Tan Son Nhut. After de-arming and reservicing, they joined other Viet-

*Aircraft 929, Sims-Harnage; 545, Beggerly-Matthews; 541, Barnett-Walker; 567, Stanley-Sjogren; and 878, Keables-McCord.

namese aircraft in a tight V-formation fly-by over the capital city at a low 300ft, ripping along at 300kts, down the main street of Saigon and over the palace. This jet bomber force was visually part of VNAF, and even more impressive to onlookers was the fact that just a few hours before, they engaged the enemy in combat, punctuating the first strike made by VNAF marked B-57s.

At Clark, the second set of four Vietnamese pilots, but having only two navigators, started their training on 22 November, and it was here that the VNAF B-57 program began to sag. By the end of the first week in January 1966, many of the aircrews in training complained of various kinds of illnesses which all but brought their training to a standstill. Further complicating matters was a training accident that occurred on 8 January 1966 when VNAF Lt Lom with an instructor pilot lost control of 838 on touchdown when making transition landings at Cubi Point NAS. The airplane became a total write off. It had arrived from the United States only 17 days before. This accident seemed to set the program back further by lowering the morale of the Vietnamese even more. Some flatly stated that they could not physically perform the maneuvers required in the B-57.

Before a solution to the problem could be reached, however, another accident occurred which had a disastrous effect on the VNAF crews and from which the program would not recover. This accident occurred on 23 February 1966, with 512 while it had Vietnamese markings and an all VNAF crew, piloted by Maj N. N. Bien. While on a strike mission, the weather turned bad at Da Nang. Because of hung ordnance and inadequate fuel remaining, Bien elected to recover at Pleiku Air Base. After de-arming and unable to start the engines due to inexperienced VNAF ground crews installing the starter cartridges improperly, the Vietnamese decided to push the airplane by hand. Bien remained in the cockpit intending to guide the airplane with wheel brakes. But hydraulic pressure had not been built up beforehand, and as the heavy airplane began to roll down a slight grade, it became uncontrollable and started off the side of the taxiway. In a freak accident, Bien jumped from the cockpit, stumbled and fell in front of the left main gear which rolled over him causing fatal injuries.

His death stunned all who knew him, for he was well liked and highly respected by both Vietnamese and Americans. Maj Bien had been a driving force and dynamic leader of the VNAF B-57 program. With his loss, there seemed to be no incentive on the part of the Vietnamese crews to stay with the Canberras. From this point on there was little or no activity by the Vietnamese in the program, and on 20 April 1966, Thirteenth Air Force terminated the VNAF B-57 operation.

16

PHAN RANG

While the two B-57 squadrons alternated their 60 day stay at Da Nang, they were attached there for operational control to the 35th Tactical Fighter Wing, an F-100 unit. As higher performance fighters took over the air war to the north, it became more advantageous to place the 366th TFW with F-4 Phantom IIs at Da Nang and relocate the F-100s. The new home for the 35th TFW became the newly established base at Phan Rang, just south of Nha Trang and Cam Ranh Bay, and the B-57 operation was moved with them. This relocation took effect on 13 October 1966, and became the last base of operation for Canberras while engaged in the Vietnam war.

Phan Rang was a place of contrast. The airfield was located on the level fertile ground which reached about three miles to the sea, while the building complex nestled at the base of rugged jungle clad mountains that extended to the west. Near the entrance to the base were ruins of a fascinating Cham Temple, reputedly built in the 12th century, while inside the base was an array of many advanced technologies. To accommodate the expanding war, this air base had been quickly built some eight miles east of Phan Rang, a city of 24,000 inhabitants.

This move initially brought about a strange operating environment for the Canberras. The new hard surface runway had not been completed at the time of this move, so a parallel AM (aluminum matting) runway was laid down beside it from which the jets were to operate. The rolling swells of the uneven surface was of lesser consequence when compared to the hazards of ingesting loose objects and dirt from the temporary surface, into the intakes of the jet engines.

At first, nearly everyone lived in tents until more suitable housing could be completed. Typical quarters became long, one story wooden structures with the upper portion of the sides left open and screened for air circulation. Eventually a number of mobile homes, commonly referred to as 'hooches', provided air conditioned comfort for field grade officers and other B-57 crews who often required daytime sleep away from the hot outdoor temperatures. Sounds of the war were always present as protecting South Korean guns would echo across 'Happy Valley' on the west side of the base, and at night flares would light the high ground known to the defenders as 'Old Smokey.'

When the engines of the B-57s started, things began to move quickly. Gear pins were pulled and stowed, and chocks were kicked away from the wheels. After the ground crewmen and the pilot had indicated that all was in readiness, the Canberras taxied out for take-off from Phan Rang on another strike against the enemy. (USAF)

Often the last action in preparing the B-57 for flight is topping off the liquid oxygen system. This can be a dangerous job due to the extreme cold temperature and explosive qualities of the liquid gas. Ed Pearson wears a protective face piece and apron during the 10-minute process. (USAF)

There was little change in combat activity for the Canberras during their stay at Phan Rang. Their daily schedule usually called for 14-18 daytime sorties and 6-8 night sorties for the B-57s. Hazards remained high and four aircraft were lost due to combat during the first 15 months at Phan Rang while the two squadrons continued their 60-day rotational stays.

The year 1967 was a golden year for the two bomb squadrons, for in May and August, the 8th and the 13th Bomb Squadrons respectively passed their 50th birthdays, they had the distinction of being the oldest continually active units in the United States Air Force. On 15 January 1968, however, the 13th Bomb Squadron, known historically as the 'Grim Reapers', was de-activated, and celebrated by flying their 1,000th combat sortie. (The 13th was reactivated later as a B-57G squadron but the linage had been broken.) This left the 8th Bomb Squadron as the only surviving tactical bomb squadron in the USAF. This called for a permanent change of station (PCS) move for the 8th to Phan Rang, leaving its ties with the 405th TFW at Clark, and becoming an integral part of the 35th TFW. This deactivation of one squadron did not effect the combat status of the B-57s in Vietnam, since just one squadron strength had been in the combat zone at any one time. It did cause another major shuffling of personnel however. Members having the longest overseas service in the squadron were returned to the US until the strength reduction to one squadron was achieved. By June 1969 the number of B-57s was reduced to nine Canberras, as 12 had been returned to the US over the previous few months to be converted into B-57Gs. Combat sorties usually consisted of a two-ship daytime strike and 8-10 single ship night sorties. A detachment of four dual control B-57Cs and instructor pilots and navigators remained at Clark to continue training the new crews scheduled to join the squadron at Phan Rang.

Two of the most difficult structural repairs occurred at this time when squadron aircraft strength was at its lowest. As a result of battle damage, a fire started in the aft fuselage section of 507, but Maj Orazio and his navigator safely recovered at Da Nang. The fate of the aircraft seemed questionable due to the severe damage, however the aft section was removed from 552 in storage at Davis Monthan AFB, Arizona, and was shipped to Da Nang for the repair.

More than once has an engine tailpipe come off a B-57 and caused major wing damage. This happened to Royce Tate in 551 when taking off on a combat strike at Phan Rang. Before an immediate landing could be made, the wing spar was severely burned, and the right wing had to be changed. Again 552 gave up a major part of herself, but the move of this wing was not as easy as the aft fuselage section had been.

The logistics involved with moving a 19ft chord wing of the Canberra is a story in itself and worth pausing a moment to recount some of the problems encountered. The first requirement occurred in 1957 when a wing at Hill AFB, Utah, was badly burned and had to be replaced. Martin built a special truck bed that would handle the wing at a 45 degree angle which afforded the best compromise to height and width. Special jack screws were incorporated that would allow the wing to slide down along the side of the truck to

The night scene around B-57s at Phan Rang was often busier than during the day. It was an around the clock job for ground crews to maintain the B-57s for continual attacks against the enemy. Maintenance technicians must share much of the credit for the combat effectiveness of the Canberras. (USAF)

Revetments were constructed soon after the arrival of the B-57s at Phan Rang Air Base. This 13th Bomb Sq B-57B has 750lb GP bombs on the wing stations. At the request of the aircrews, the standard light gray undersurfaces were painted flat black for better concealment during night missions.

further reduce the height for passage through an underpass. If preferred, it could even pivot to lay flat where with clearance permitted it to pass under an obstacle.

In addition to the engineering and many man-hours to build such a rig, nearly three months were necessary for obtaining wide-load clearances from all the states between the factory in Maryland to Utah, through which the wing was to be transported. The clearances added to the expenses as well as paying each state for police escort all the way across the many states.

When the appointed day to begin the highway trek arrived, the wing moved no further than three blocks from the factory before a car ran an intersection and rammed the wing broadside, rendering it useless. To prepare a replacement wing for this trip, this prolonged the schedule beyond existing limits that the movement by highway was finally abandoned.

Cutting off portions of the wing to reduce it in size for loading into a C-124 was considered. Wind tunnel test on models for possibly carrying the wing on the bottom of a C-124 was also examined, none meeting with success. The wing was a *monster* to move, in the literal sense. Eventually it was learned that the Southern Railroad had a special rail car designed to carry oversized transformers. With modifications, it was able to carry three B-57 wings, although special routing was required for reaching needed places such as Hill AFB at Ogden, Biggs AFB at El Paso, and San Diego, for shipment to Vietnam.

It was said that the Australians solved the same problem by building a special truck with low pressure tires that departed the highway system and moved directly across the open desert to the destination. This weird contraption gave startled jack rabbits the *hoo hoos*!

New arrivals and freshly camouflaged B-57s stand initially in open lines at Phan Rang AB until revetments can be finished. This was a major change from the crowded conditions from which the B-57s operated at previous bases. At Phan Rang, the on station squadron of B-57s was assigned to the 35th TFW, a F-100 outfit. (USAF)

Nose art was not as popular during the Vietnam war as it was during World War II. Some evidence was often found, however, and B-57s were no exception. While at Phan Rang, these characters appeared from unknown hands, perhaps applied with magic-markers.

Left and Above: No wonder that Dick Burkholder, pilot (left) and Milton Stein, navigator (right) look happy. They were able to recover safely at Ubon, Thailand, after anti-aircraft fire knocked off most of the right stabilizer and all of the elevator on 282. The damage occurred during a strike in Laos on 4 April 1969. (E. Youngs)

But the wing did finally reach Phan Rang after months of expectant waiting. 'I was expecting the wing to arrive on a day to day basis,' reported Gene Youngs, who was Martin's Senior Systems Engineer at Phan Rang for the B-57s. 'I saw a truck squeeze through the gate, carrying what I thought was a Vietnamese *house* – not realizing until after I returned from lunch that this was a crate containing the long awaited spare wing. The container was *huge!*'

An unusual combat loss took place on the night of 13 December 1968 on what started out to be a routine strike over Laos for Majs Dugan and McGoldrick. During their ground attacks in the vicinity of Xieng Khovang, their B-57E, 284 collided in the dark with Candlestick, a C-123 flare-ship with whom they were working, and all crewmembers perished except the C-123 co-pilot.

A lone B-57 of the 8th Bomb Squadron returning from a mission circles its home base of Phan Rang. This was the east side of the base from which mostly transport aircraft operated. (USAF)

This familiar view to Phan Rang based Canberra crews shows clearly the B-57 parking revetments just above the airplane. F-100s of the 35th TFW fill the smaller revetments below. The main base with housing facilities is the area to the right. (USAF)

Before long, the end of operations was in sight for the B-57s that had fought the war for such an extended period. In late September 1969, the Canberras and their crews began leaving Phan Rang and 'Happy Valley' for the last time, recovering at Clark to prepare for the long flight to the United States. Frustrating as it was, Dudley Mizer stayed behind to fly the last B-57 out of Vietnam, that being 551 after having its wing replaced. Dudley was the maintenance officer for the 8th, yet creating history as pilot of 'the last American bomber to leave Vietnam' was the least of his concerns when he departed Phan Rang on 15 October. (For being 'the last' gave reason to select 551 for inclusion in the National Air and Space Museum collection.)

While this movement of airplanes was taking place, the designation of the 8th was transferred to a unit at Bien Hoa AB, to become the 8th Attack Squadron, equipped with Cessna A-37s, and the long tradition of the 'Liberty Squadron' continued.

This brief account of B-57 activities at Pan Rang is disproportionate for the three long years that many combat strikes were flown from the base. Operations had stabilized however, and aside from losses, the war was being fought without the upheavals experienced in the earlier years in Vietnam. This period did include the Tet Of-

The last jet bomber to leave Vietnam as a result of the Vietnamization Program, was 551, which departed in November 1969. It was assigned to the 8th Bomb Squadron, although this earlier picture shows 13th BS markings. This airplane now resides with the National Air and Space Museum. (R. Walker)

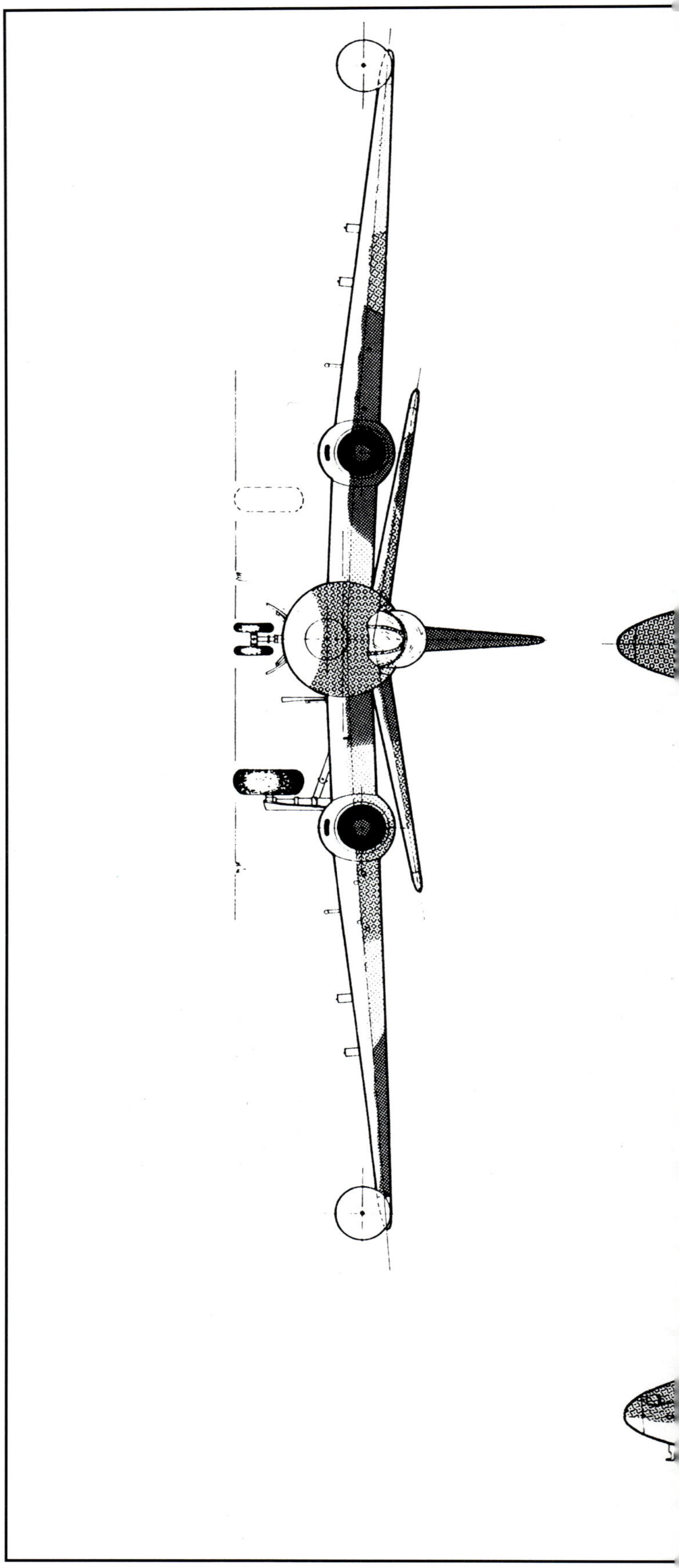

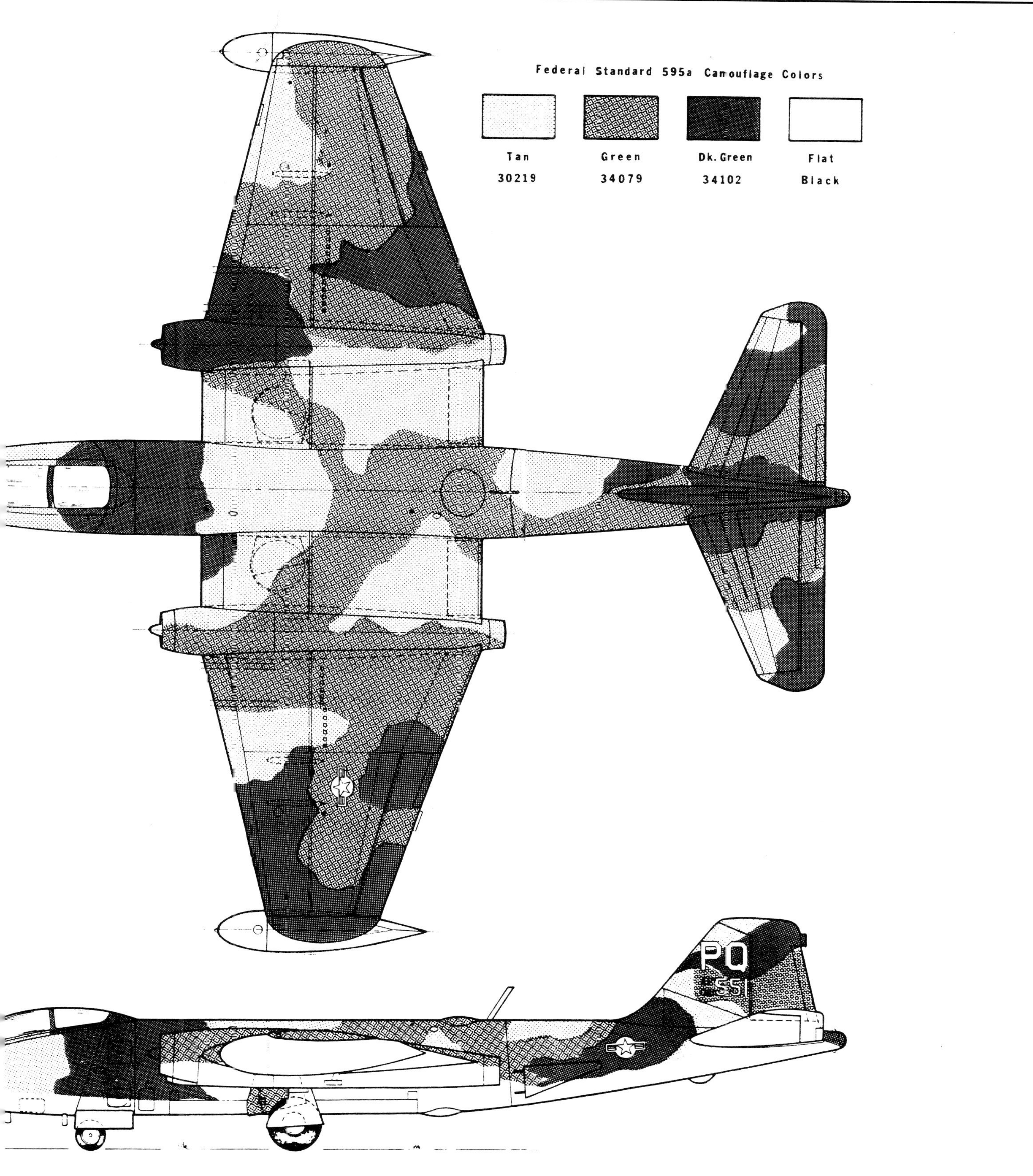
Federal Standard 595a Camouflage Colors
Tan
30219
Green
34079
Dk. Green
34102
Flat
Black
PQ
551

The early morning sun streams into the revetments that house B-57s at Phan Rang, as ground crews busy themselves for the days bombing missions. At Phan Rang, it was an around the clock mission in launching and recovering the Canberras, and preparing them to go again in a matter of a few hours.

fensive of February 1968, a time that 22 Canberras were often supporting 30 sorties a day to counter these pressing attacks. When the offensive was turned around, the Canberras and their crews continued hitting the enemy and hurting him badly over the entire period, never letting up until the last day they were fragged for combat sorties.

One interesting statistical note here is that of the 94 B-57s that were assigned to these two bomb squadrons during the six year wartime period in SEA, 51 were combat losses (including 15 destroyed on the ground). Of the original 47 deployed from Japan, only one (53-3879) remained to the end, however 11 were withdrawn early to support the B-57G program, 27 withdrawals for this program in all. When 31 October 1969 arrived and the B-57 operation of the 8th Bomb Squadron was ended at Phan Rang, 13 aircraft were on hand. (See Appendix 14.) These airplanes went into short-term storage at MASDC at Davis-Monthan AFB, Arizona, from which a number were regenerated to operate with DSES units. This concluding mission for B-57s is described later in this book as the 'Friendly Enemy.'

Two more on the way! Ground crewmen have assisted engine starts for 898 and 567 as they leave their parking areas at Phan Rang. B-57 operations were a day and night affair, keeping the VC on guard against the lethal and relentless bombs from these Canberras. (USAF)

17

THE FAC's POINT OF VIEW

'Give me four A-1s (Douglas Skyraiders) or one B-57', was an often quoted request for strike aircraft made by FACs (Forward Air Controller), for the B-57 was the A-1 of the jets for effective pinpoint bombing – and then some.

The foregoing accounts about the B-57s in Southeast Asia, retell the experiences of others, plus the notes that I kept while history was being created. For this portion on the FAC's viewpoint about B-57s, I can speak even more authoritatively, for I was a FAC, and I directed strikes for B-57s during 1967/68. To keep in closer touch with my B-57 buddies, I often made stopovers at Phan Rang to hear first hand how the war was going for them. I had been separated from flying Canberras having remained in Japan for another assignment when the two squadrons moved to Clark. As the war continued, my number came up for a FAC assignment before it appeared on the B-57 assignment list – much to my surprise and disappointment.

Admittedly, the FAC assignment turned out to be interesting and challenging. Under the rules of engagement used in Southeast Asia, ordnance could not be expended in South Vietnam and certain areas of Laos without a FAC to control the strike. Being in a low, slow flying observation plane, it was the FAC's responsibility to locate the target before ordnance was released and ensure that the attack was aimed at the correct target. This all seemed simple enough except for this last requirement when working with the higher performance aircraft. A FAC could readily tell what kind of a day he was going to have by the type of strike aircraft assigned him for preplanned targets. What was hoped for (in alphabetical order) were A-4 Skyhawks, A-37 Tweetybirds or B-57s. As far as jet aircraft were concerned, I found these three types to be the most effective air to ground support aircraft available at that time. The higher performance aircraft needed more room in which to work and several types which will remain nameless – would hardly come down low enough to the target whereby I could see them when they called 'on target' so that I (the FAC) could respond with 'cleared to drop'.

The B-57s were never a problem. Normally their pattern altitude was a comfortable 7,000-8,000ft AGL, which kept them always in sight of the FAC that worked near the ground, and he could see all the other members of the flight in the same pattern. When a FAC cleared a Canberra to bomb or strafe, its arrow-like fuselage readily pointed at the target, leaving no doubt in the FACs mind as to what it was lined up with. In a turn, the broad wing would show up well, silhouetted against the sky, which helped to keep track of all the Canberras in the attack pattern.

Working with B-57s at night was still another type of experience. I recall one night while operating in southern Laos, I was having a very difficult time locating road traffic on which to direct three flights of aircraft that were to work with me. I was running a little behind time that night in my relatively slow moving Cessna O-2A, and I hardly had enough time to reach the target area and

Air strikes in South Vietnam and many areas of Laos had to be controlled by Forward Air Controllers (FAC). Their aircraft were normally the single engine Cessna O-1 Bird Dog, or the twin-engine push-pull Cessna O-2A Skymaster. The latter became operational beginning in 1967 for out-of-country and night missions.

The low, slow flying FAC could more readily spot a target than a crew in a faster moving jet. The FAC in this Cessna O-2A prepares to mark the target with smoke rockets, then to direct the air strike for B-57s or fighter-bombers on to the target in relation to the smoke marker.

A two letter unit marking system for aircraft carrying SEA camouflage was started in 1967. Seven of these letter codes were assigned to B-57 units and a sample of each is shown here.

PQ, 8th Tactical Bombardment Squadron, 405th Tactical Fighter Wing (Clark AB), and later assigned to the 35th TFW (Phan Rang). Squadron color: yellow.

PV, 13th TBS, 405th TFW (Clark AB), color: red.

FK, 13th Bomb Squadron (Tactical), reactivated and first assigned to 15th TFW and carried this code to 8th TFW, Ubon, Thailand, when equipped with B-57Gs.

FS, 4424th Combat Crew Training Squadron, 15th TFW, later TFW, MacDill AFB, Florida, for training B-57G crews. Black on rudder tip.

JM, 4416 Tactical Electronics Warfare Squadron, 363th TRW, Shaw AFB, SC, with two EB-57Es. Top of rudder was gray.

JO, 22th TRS, later 62th TRS, 363th TRW, Shaw AFB, lost one B-57E (269) that crashed. Tactical Air Command insignia. Serial number shown on JO and JM is incorrect. Should be 54274. White letters are 8in x 24in. Numbers are 4in x 6in and 10in x 15in. (D. Anderton)

GT, 556th Reconnaissance Squadron, 347th TFW, Yokota AB, Japan with four EB-57Es. Squadron later moved to Kadena AB, Okinawa, and B-57s eventually joined the 18th TFW. (R. Walker)

begin sorting out the few dim lights as being trucks before the first attack aircraft were to arrive. The first to call in was a B-57 crew. I explained my predicament, and they willingly set up a holding pattern to wait until I had something for them. Moments later, a flight of F-4s reported in, already short of fuel, and had to drop quickly. I worked them in on the best target I could locate at the moment. I marked the target with smoke rockets then dropped a flare. The Phantoms made one pass each, dropping everything, and then they were gone.

Relieved that they were safely away, I started checking further up the valley for additional targets, having plenty of time until the third flight was to arrive. Fifteen to 20 minutes could have passed when, 'Covey 43, this is Yellow Bird 48 – do you have any targets for us yet?' I had totally forgotten I had friends waiting, for in the meantime, I had heard nothing from them. B-57 crews seldom pressured the FAC, for they seemed always to have enough fuel to do the job whenever you might be ready for them. Some pilots of other aircraft had to be on the radio to the FAC or others in the flight all the time, but not the case with B-57 crews (perhaps because they had a back seat member to talk to on intercom!).

Working with a B-57 at night was an experience. Our Cessna O-2A Skymasters carried two flares on the wing and 14 smoke rockets. With a good target such as an active river ford, the attack was started by 'hosing off' two smoke rockets in a dive when aligned as best as possible to where you *believed* you last saw the target. With a sharp pull-up and a climb for altitude to 3 or 4,000ft, a wing flare was then dropped to light the target. Informed on timing, the B-57 crew would set up a pattern and were ready to attack as soon as the flare ignited and the FAC called out the target in relation to the two puffs of white phosphorous smoke from the rockets.

The flare did not last long, but by the time it burned out, the Canberra would have started sufficient fires that kept the target well marked. One weapon called the 'Funny Bomb' was spectacular. These were M35s and M36s that were a large canister that opened when dropped. The contents spread as tongues of fire before reaching the ground which seemed to persist for 10-15 minutes and often longer due to sustained fires that they started. Watching from the air it appeared as a blanket of glowing coals being spread uniformly over a large area as they floated to the ground. Anything they touched that would burn – did. Seemingly, only B-57s carried these weapons, yet they were very effective for area coverage.

Many statistics evolved from the air war in Southeast Asia, such as the number of bombs dropped per each type of aircraft, loss rates, etc, but there was no scale by which to grade the effectiveness of each aircraft type in attacking ground targets. The FAC was in the best position to observe these results and could compare them with other aircraft doing the same job. Consistently, the B-57 would be named as a top performer. Little was said about this in the news media, however, as focus was always placed on the latest types of equipment being employed. The reason was clear, for the military was desperately in need of new equipment, and to emphasis the effectiveness of the near obsolete type of aircraft was no way to influence the money holders for getting new appropriations.

(See Appendix 10 for Combat Comparison figures.)

18

RAAF Canberras

It would be improper to ignore the Canberras flown by No 2 Squadron of the Royal Australian Air Force, with all that has been said about the American B-57 activities in Southeast Asia. Like the B-57 units, it too was stationed at Phan Rang and came under the operational control of the 35th Tactical Fighter Wing as were the American squadrons when on rotation at that base.

The Australian-built Canberra Mk 20s arrived at Phan Rang in April 1967 for a stay, unknown at the time, that would last over four years and extend beyond the time that their American counterparts would be moved back to the United States. This squadron was the third and last operational RAAF unit deployed to South Vietnam, the other two units being equipped with Caribou transports and Iroquois helicopters.

At the beginning of their operational assignment, the squadron of eight Canberras was assigned exclusively to night attack, making drops normally from 20,000ft, directed by ground radar. These

The early morning haze covers South Vietnam as this Australian Canberra heads toward its target. One squadron of these Australian-built Canberras occupied Phan Rang AB along with the American B-57 squadron for nearly four years. (USAF)

Resting in the early morning sunlight after ending its bombing mission just before daylight, an RAAF Canberra awaits its ground crew to prepare it for the next night's mission. The Mk.20s were the only level bombers based in Vietnam able to employ a bombardier and a bomb sight.

operations were known as *Combat Sky Spot* strikes, which achieved amazing accuracy on fixed targets. They were not intended to replace visual bombing for which the Canberras were configured, but to complement it during poor weather and darkness when visual methods were ruled out.

It was mutually agreed by 7th Air Force and No 2 Squadron, that RAAF bombers would launch eight sorties a day against up to 16 preselected targets, every day of the week. These targets were to be selected by 7th Air Force, and would be anywhere within the four corps areas of South Vietnam. For their radio call sign, the name 'Magpie' was appropriately selected, which symbolized that unit by being a derivative of their squadron insignia.

The first day of operation was 23 April 1967, which called for eight strikes with the aircraft taking off at hourly intervals. They dropped 42 bombs of 500lb each on 11 targets ranging from II Corps in central Vietnam to IV Corps in the Mekong Delta region in the far south. As with all missions of this type where bombing results could not be assessed, the only satisfaction to the crews was that their mission was flown as planned. The Canberras had quickly proven themselves competent in radar-controlled bombing at night,

Preparing to drop its bomb load through low clouds on the enemy, RAAF Canberra Mk.20 of No 2 Squadron is on a 'Combat Sky Spot' mission, where direction and bomb release is calculated by ground radar. This method was frequently used at night by RAAF Canberras and occasionally by USAF B-57s. On the wing tip of A84-231 is a 750lb GP bomb.

Blue tail flashes identify Canberras of No 2 Squadron. Miniaturized national insignia follow the same theme as other camouflaged aircraft operating in SEA.

External bomb racks for the Mk.20 were noticeably different and more restrictive than that of the B-57s. Shown here is a 500lb bomb on a wing tip rack, as there were no attachment points for bombs on the main wing section.

but most RAAF Canberra crewmembers had been trained in visual bombing by day and felt that in this role the Canberra potential was much greater. They therefore supported proposals for having the Canberras put to use in this type of operation.

While USAF squadrons were carrying out day missions under the control of a FAC, the RAAF crews were eager to take part in the same activity. Authorities, though, considered the RAAF Canberras unable to add significantly to the work already being done by the American tactical aircraft since the Canberra Mk 20s were configured for high level bombing. However, as early as 25 June 1967, the Canberras successfully demonstrated that the aircraft was suitable for this type of weapon delivery. Beginning in September 1967, bombing under FAC control began and was immediately successful. Thus, No 2 Squadron became the only squadron in South Vietnam employing a level bombing technique using a precision bombsight of World War II vintage.

Starting from 10 November 1967, up to 50 percent of the squadron's effort was given over to daylight bombing throughout

During its near four year operation in Vietnam, No 2 Squadron lost two Canberras. The crew of one was safely recovered while no trace was ever found of the other.

Vietnam under FAC control, which became known as 'Booma' Missions. On 19 November, two Canberras carried out a successful strike during heavy fighting around Dak To in the II Corps area. Dak To, was only a few miles from the Laotian and Cambodian borders and the enemy took advantage of the nearby border sanctuaries to strike with speed and strength. Four North Vietnamese regiments were sent into action. The Allies quickly reinforced Dak To with American and South Vietnamese army units and sent in an array of tactical air strikes together with mass B-52 missions. On that day, Wg Cdrs Aronsen and Hughes had been briefed to carry out a radar bombing mission in another area. But this mission had been cancelled and the Canberra crews, reluctant to return to Phan Rang without having dropped their bombs on a target, sought alternatives. Eventually they received a call for an urgent task at Dak To, much further to the north. They agreed to do the job knowing they could land at nearby Pleiku AB for refuelling. At Dak To the FAC briefed them to bomb enemy troops who were firing from one side of an 8,000ft ridge on friendly forces on the other side. They were using three heavy mortars and four heavy machine guns. There was little margin for error, but after the FAC had marked the target the Canberras bombed right on his smoke marker. Back at Pleiku where they landed for reservicing, they met the FAC who told them their bombing had been 'magnificent.' He said that after the Canberras had left, the friendly forces, who had been taking severe casualties from the enemy fire, did not hear as much as a 'peep' from the hostile guns and mortars.

By 1969 about 70% of No 2 Squadron's missions were being flown in the Mekong Delta area where the Canberra's characteristics were best suited. The Canberras gave the best results in the delta because of the region's flatness and the fact that the altitude of the target which was almost always just a few feet above sea level could be fed into the bombsight with precision.

Canberras of No 2 Squadron had received battle damage from small arms fire and debris from their own bombs by flying too low during attacks. It was not until 3 November 1970, however, three and a half years after it began bombing in Vietnam, that their first Canberra and crew were lost. This was Magpie 91, assigned to a 'Combat Sky Spot' mission in the Da Nang area. The radar bombing operator reported good radio contact with the aircraft. Plt Off Robert Carver released the bombs from a height of 22,000ft at 20:22hrs following a normal bombing run, after which the Canberra's pilot, Flt Off Michael Herbert, reported that he was turning to a heading of 120deg. This was the last transmission from the crew of two aboard A84-231, and after an extensive search there was never a trace of the missing Canberra.

The second and last RAAF Canberra was lost on 14 March 1971, in the north-west corner of South Vietnam. Fortunately this crew was recovered after one harrowing night's experience, both being separated at this time, listening to the sounds of the jungle where tigers, elephants and many other wild animals and reptiles roamed. Wg Cdr J. Downing, CO of No 2 Squadron and Flt Lt A. Pinches, his navigator, had the dubious honor of being the only RAAF crew shot down by a SAM missile. An immediate result of the loss by surface-to-air missile of the RAAF Canberra was that 7th Air Force directed that no aircraft that was not fitted with SAM detecting equipment would be permitted to operate within the range of known SAM sites. More and more, this lethal enemy equipment was moving into South Vietnam and soon all of the I Corps area (north quarter of South Vietnam) was no place for aircraft to fly without SAM sensing equipment. By this time, the American B-57s had been withdrawn from the war through the Vietnamization Program, therefore it came as no surprise when the RAAF squadron was also to be sent home. The 35th TFW was scheduled to leave Vietnam and without their operational and logistical support, their stay at Phan Rang would have meant endless problems. As a result, the squadron flew their 11,963rd and last sortie on 31 May 1971. Long range ferry tanks were then installed and the Australian Canberras left Phan Rang on 4 June for Amberly, via Darwin, to begin a new assignment with their airplanes as a reconnaissance and target towing squadron.

19

Patricia Lynn

There were so many classified programs connected with the 'Patricia Lynn' Project, that when the name was mentioned, it was often done hesitantly and in a lowered voice. The aircraft for this USAF project were reconnaissance RB-57Es, stationed at Tan Son Nhut. Since these were the first aircraft equipped with IR (infrared) sensors to be used in-country, they were under very tight security control. Not only were these the first jets to be introduced into the air war in Vietnam – May 1963, they became the last of the Canberras to leave Vietnam. What was intended as a very short duration project, became the most permanent – lasting over eight continuous years, the longest of any jet aircraft in that conflict.

The need for a more sophisticated means of gathering aerial intelligence became increasingly critical as U.S. military became more involved in Southeast Asia. The U.S. Air Force had developed several new reconnaissance systems, and where better to service test them than in Vietnam. General Dynamics plant at Fort Worth was awarded the contract to modify two former tow target Canberras to carry this new equipment. Initial modifications included the redesign of the forward nose section to house a KA-1, 36in forward oblique, and a low panoramic KA-56 camera. Mounted on the inside of the specially configured bomb bay door was a KA-1 vertical camera, K-477 split vertical day-night camera, and infrared scanner, and a KA-1 left oblique camera. So urgent was the need for reconnaissance information, that on arrival at Tan Son Nhut, the ferry crews were *now* combat crews and were immediately briefed by 2nd Air Division staff for missions needed to be flown. The day after their arrival, Capt Bill Scott as pilot, with navigator Lt Bill Sung, flew the first mission on 7 May 1963, in 243. Three days later on 10 May, Capt Don Wachholz and Lt Leo Otway completed their first mission in 245. Infrared photography with good

'Patricia Lynn' RB-57E aircraft were easily distinguished by their uniquely shaped nose. Note camera window. Here 243 starts-up during initial ferry flight to Vietnam in May 1963. Aircraft remained unpainted until 1965. (D. Wachholtz)

This 'Patricia Lynn' Canberra survived only two years of the war. Although 243 was the first of six performing this mission, small arms fire during a night reconnaissance mission brought it down in August 1965, but it was able to return its crew to just short of Tan Son Nhut to parachute safely. (USAF)

Cramped conditions are apparent in this view in which Patricia Lynn had to park and operate. The tails of three Pat Lynn RB-57Es are shown in this view taken from a transit VNAF B-57B, with another VNAF Canberra on the other side of the jet blast shield. Picture was taken 29 October 1965.

This is an early view of 249 before receiving an all black finish while operating from Tan Son Nhut. Manufactured for the benign role of target towing, B-57Es were used exclusively in the "Patricia Lynn" mission as well as combat replacements for the 8th and 13th Bomb Squadrons. (R. Schultz)

photo interpreters identified such things as VC base camps, small arms factories, storage and training areas that were not otherwise detected by the naked eye. Results of these missions were impressive and the 'Patricia Lynn' Project aircraft were kept busy from that point on.

By July 1963, the TDY crews returned to their 6091st Reconnaissance Squadron in Japan and were replaced by permanently assigned crews fresh from the States. The unit became Detachment 1, of the 33rd Tactical Group, and continued to grow, receiving two more similarly equipped RB-57Es, 237 and 249. These two aircraft were flown from the General Dynamics Plant in Fort Worth where they were modified, then flown to Vietnam in December 1964. Their crews, adding further to the strength of the unit were Bill Ingeman and Jerry Herts in one and Roy Marsden and Bob Nelson in the other. The first of these airplanes to be in the black finish was 264 which arrived in that scheme at Tan Son Nhut on 22 November, 1965. This established the maximum number of five aircraft for this unit. More often referred to as just 'Det 1' throughout their operational period at Tan Son Nhut, their parent unit became the 6250th Combat Support Group on 8 July, 1965. A year later they were assigned to the 460th Tactical Reconnaissance Wing which lasted to the end of their service. In the air, their radio call sign 'Moonglow' readily identified these Canberras.

As early as 1965, Patricia Lynn crews were flying BDA (bomb damage assessment) missions into Laos teamed with B-57 bombers and a C-130 flare ship as already described in 'Night Intruder Missions.' Bob Schultz and Tom McNeillie were the first to fly these night photo missions for assessing the damage along the Ho Chi Minh trail, better known as 'Steel Tiger' missions. A BDA mission 'up north' with F-100s was flown in mid-March 1965.

Combat situations produce losses, and this unarmed low-flying reconnaissance operation was no exception. The first within a unit is always the most memorable.

While engaged in an in-country night infrared reconnaissance mission on 5 August 1965, 243 received hits from small arms ground fire. The degree of damage was not known at the moment, but circuit breakers began to jump out in rapid succession. Home base was close at hand and the situation seemed not too critical at that point. Aileron control finally gave out, indicating a possible fire in the bomb bay. Unknown to the crew, fire was seen by others to be

"Got cha!" To disprove a claim made by a Patricia Lynn pilot that no one could get on his tail without him knowing it, this picture was taken by another "Pat Lynn" pilot to make a point. The nose-mounted camera took this view of 249. (R. Schultz)

Canberras took their punishment on the ground as well as in the air. Hundreds of patches cover holes on 264, inflicted during a mortar and rocket attack on 14 April 1966 at Tan Son Nhut. (USAF)

streaming along the left side of the fuselage. The pilot, Capt Dick Damon, gave the order to eject and both crew members parachuted to the ground safely. The plane crashed a mile and a half from Tan Son Nhut.

The humor of this accident lies in the fact that navigator Dick Crist landed in the security of a South Vietnamese (ARVN) training camp. Dick did not know this, so stayed hidden in the tall grass in which he landed. Not only had his chute just opened a moment before he hit the ground, but his seat caught on the shroud lines and came down on his head. Fortunately he still had his helmet in place. The matted grass cushioned his impact that could otherwise have been fatal.

In the darkness Crist was suddenly aware that a Vietnamese soldier had walked to within one foot of him and had yet to see him. This friendly ARVN soldier and others were looking for Dick to give help. But Dick, expecting this could be an enemy soldier, uncoiled his tall and husky body, nearly 6' 4", and coming off the ground, he grabbed the Vietnamese around the neck and shoulders with one hand in a near death-grip and pressed his .38 pistol to the side of the soldier's head with the other. The slight framed fellow began kicking and thrashing because his feet couldn't reach the ground. Afraid this giant American would kill him, he kept yelling, "No VC! No VC!" At that moment, Crist became aware of many others surrounding him as his eyes grew accustomed to the darkness. It was now apparent that he was among friendly South Vietnamese and he hastily withdrew his weapon. This story was often told among flight crews, and it aptly illustrated that a quick trigger finger could have suddenly caused a *friendly* group to become *very unfriendly*!

The second and last RB-57E, 264 was lost on 25 October 1968 after being hit by ground fire in the left engine. Both Capt J. J. Johnson and Maj Phil Walker ejected and were recovered safely.

En route to Gifu AB, Japan, for further repairs, 264 is seen here on 26 June 1966 departing Itazuke AB after refuelling stop. (H. Inoue)

The presence of 'Patricia Lynn' RB-57Es were acknowledged throughout the Vietnam war, but their actions were seldom openly discussed. Even their black painted exteriors make them look sinister and crews felt that this unusual decor attracted more than their share of enemy fire. (L. Mongeon)

Fully repaired and repainted, 264 is again ready for action. RB-57Es of the 'Patricia Lynn' project established a remarkable record for any type of aircraft. Over this eight year period two of the six airplanes were lost to combat, and two were with the project from beginning to end. One flew over 8,000 hours and retired to the U.S. to fly in a more peaceful occupation. (USAF)

Trips between Vietnam and the United States were not uncommon for Patricia Lynn aircraft because of the many modifications made in upgrading their equipment. Shown here is '249 on a stop-over at Elmendorf AFB, Alaska. (B. Jones via D. Menard)

This aircraft was replaced by a sixth Patricia Lynn aircraft, 55-4257, this one equipped with (TFR) Terrain Following Radar. Several of the newly assigned pilots, including the commander designate Roy Marsden, Bill Ingeman, and Bob Schultz, were given flight training in November 1964 with this special equipment before leaving the States. They found its use to be very effective for target coverage but very hungry in using up IR film. Unfortunately, as flight crews rotated and new ones did not have this specialized training, they did not trust the TFR, so it received little use in later years.

Bob Schults, an early user of the Terrain Following Radar in Vietnam, had very positive things to say about the equipment. He conveys his point of view this way. "I had considerable confidence in the TFR, and flew it at every opportunity to keep in practice. The installation of the TFR equipment in our aircraft was for the purpose of allowing the crew to maintain a constant altitude while flying infrared missions and thereby insuring the best imagery. Unfortunately, theory and application did not coincide. The TFR equipment had to be manually set on the ground for either 500 or 1,000 feet. At these altitudes, it was impossible to get any reasonable target coverage before running out of IR film, since most infrared missions were area covers attempting to locate VC and North Vietnamese troop concentrations." Bob continued by saying that, "Flying at a low altitude did provide excellent IR imagery, but it was not possible to completely cover more than one target. Through experience, it was agreed that the best altitude to obtain useable imagery and cover the most targets was 2,000 feet."

There were frequent changes and up-dating of the equipment carried aboard these Canberras, often requiring them to return to the United States for modification. The last modifications made in the late 1960s put 12in focal length KA-82, and 24in focal length KA-83 cameras into the aircraft. These were the first tactical uses of high-acuity, high-resolution photo systems originally designed for spy satellites. The film required special processing but the results were rewarding.

New intelligence gathering equipment test projects carried such names as 'Compass Haste', 'Compass Sight', to name a few. In 1968, 'Compass Eagle' was perhaps the most significant project of them all. Gerry Reponen who was a 'Patricia Lynn' Project pilot at the time, described this equipment as an infra-red scanner and an in-flight display screen. There was no need for picture taking and having to identify targets after being developed on the ground. The display screen showed immediately what was happening in the darkness below. This was particularly useful over the rivers southeast of Saigon where the enemy moved supplies at night in sampans. Gerry flew these sorties they called 'Moonriver Missions', and on the first night they located a fleet of VC sampans. By prearrangement, the Navy was notified by radio of their exact location, enabling them to dispatch PBRs, helicopter gunships or artillery fire into the area. Through this watchful eye, river traffic slowed considerably.

In 1969/70, 'Patricia Lynn' missions flown regularly in Laos ('Steel Tiger' and 'Barrel Roll') were renewed. There were also special missions into Cambodia to support the U.S. invasion in 1970. Det 1 received a special citation from MACV in 1970, stating that the RB-57s had provided over 94% of the battlefield intelligence. These few Canberras acquired more daily targets, both day and night, than the two RF-4C and one RF-101 squadrons of the 460th TRW combined. Statistics showed that with four more RB-57Es similarly equipped, the 'Patricia Lynn' aircraft could have accomplished all the day and night in-country reconnaissance during the Vietnam War. This was attributed to the Canberra's stability, maneuverability, versatility of equipment, a longer time over the target area, and not to be overlooked – crew ability. Except with 'Compass Eagle' late in the war, crews relied only on visual map reading to locate, identify, and acquire the various types of targets which included night operations. Although these missions were flown at low level, often in darkness in mountainous terrain and too often in marginal weather, they were done without internal radar or outside navigational assistance. When the 'Patricia Lynn' operation was terminated after mid-1971, one (245) or more of the RB-57Es nearly reached the 8,000 flying hour mark.

20

B-57G Night Intruders

During the war in Vietnam, there evolved a major modification to the bomber version of the B-57. Originally conceived as a night intruder, it could do the job no better than the earlier Douglas B-26 which it replaced when called upon to perform night missions in Southeast Asia. The B-57G was developed to fill the requirement for a completely self contained night attack system specifically designed to interrupt the flow of enemy supply traffic over the trails of Laos. Thus, 25 years after it was acquired as a night intruder aircraft, the B-57 was equipped to fill its mission properly.

To provide night intruder configured airplanes, modifications were made to existing B-57Bs, which then became B-57Gs, a creation of the project known as 'Tropic Moon III'. The earlier 'Tropic Moon II' program experimented and combat operated from Phan Rang AB from December 1967 until July 1968, with a low light level television mounted in a pod under the left wing of three B-57Bs, 518, 580, and 860.

By June 1969, 16 Canberras had been withdrawn from Phan Rang and returned to Baltimore, Maryland (see Appendix 6). There, under a joint contract, Martin modified the nose sections of these aircraft, and Westinghouse, the prime contractor, installed newly developed sensing and tracking systems. While these aircraft were in the various stages of completion, the first 'G' model accepted by the Air Force was flown from the factory in May 1970 by Lt Col Paul R. Pitt, to MacDill AFB, Florida. Pitt was the newly appointed Commander of the 13th Bombardment Squadron, Tactical, that was reactivated on 8 February that year at MacDill. Lt Col Charles R. (Chuck) Strain came aboard shortly thereafter as Operations Officer. After intensive and accelerated training, the 13th Bomb Squadron deployed to Ubon, Thailand, in September 1970, and became part of the 8th Tactical Fighter Wing with 11 B-57Gs. Follow-on training for replacement crews was handled by the 4424th Combat Crew Training Squadron at MacDill which retained four of the 'Gs' and the four 'Cs'. One B-57G, 905, was lost the previous December in an unfortunate fatal accident during the single engine test phase while being flown by Martin test pilot Bob Turner.

The newly equipped airplanes contained three sensors; forward looking radar, infra-red and low light level television plus a lasering device that was originally designed for ranging only. The systems operator worked the equipment from the rear seat while the information flowed through a computer for automatic bomb drop after receiving consent from a switch located in the front pilot's com-

To attempt again to develop a true night intruder bomber, the B-57G model of the Canberra series evolved. Airframes of 16 B-57Bs were withdrawn from combat and extensively modified to incorporate among many things, electronic sensors for night and all weather attacks. Westinghouse and Martin made these modifications.

There was no mistaking the B-57G with its enlarged, jowl-like nose section which housed the low light level television system and other sensors for detecting moving ground targets and controlling laser guided bombs. Forward pointing boom contains pitot tube. (E. Youngs)

Resting in the hot sun at Ubon AB in December 1970, this B-57G is being readied for another night mission along the Ho Chi Minh Trail in Laos. Modifications to these 16 Canberras cost about $49 million, an extremely expensive program which was often criticized. The operation ended after two years of combat. (USAF)

partment. The pilot received direction from the Flight Director System for target alignment. Bombing runs were flown at 4,000 to 6,000ft above the ground.

A technique learned during the latter phase of combat crew training while still at MacDill AFB, was that the ranging laser could be set to deliver the 10 pulses per second required to designate targets for laser guided bombs. Since bombs always impact behind the delivering aircraft, yet this sensor could only look down a maximum of 89-1/2 degrees, it was necessary to develop a way for the sensor operator to keep the ranging laser beam on the target until impact. This was accomplished by beginning a 2G pull-up at weapon release, maintaining this through a 500ft gain in altitude, then reversing to a 0G pushover until bomb impact. The B-57G thus became the first aircraft to drop the weapon and designate the target

Redeployment of the 13th Bomb Squadron to the war in SEA with B-57Gs went without incident. Raising the canopy of his Canberra at Kadena AB en route to Ubon AB, Thailand is Darrell Van Citters with navigator Ron Reppe. Others look on with awe at this strange – yet sinister-looking war machine. (P. Pitt)

B-57G night intruder bomber; reconfigured from existing B-57Bs, with redesigned nose to house three sensors (forward-looking radar, infra-red and low-light-level television) plus a lasering device; hydraulically-driven generators added for more electrical power; 16 reconfigured. (Courtesy of Aerospace Publishing, Ltd.)

from the same aircraft. Once a target was identified, one of four 500lb laser guided Mk 82 'Smart Bombs' carried on the wing stations would be released. Four M35 or M36 'Funny Bombs' were carried in the bomb bay for use with the censored bombing systems. When these were exhausted from stocks, M117 750lb bombs became the standard internal ordnance. The former wing guns had been removed, for these modified Canberras were exclusively level bombers.

From the early 'Tropic Moon' days that led to the development of the 'G', Paul Pitt had been with the program and had many good things to say about the airplane. The 'G' had a new instrument landing system for better all weather capability. With the docile flying 'feel' caused by the enlarged nose, the airplane was very stable which made instrument landing quite easy. It was a bear, however, when landing with one engine and it was wise to hold 10kts more than the minimum single engine airspeed of 155kts for the standard B-57. Some said that it was under-powered with the added frontal area, and during the redesign phase the installation of a more powerful engine was considered. But speed was not essential according to Paul, for this mission was normally flown best at between 250 and 300kts indicated.

Staying with the Canberra from the very beginning was Gene Youngs, Martin Senior Systems Engineer. Youngs, known as 'Mr B-57' to those that admired his technical knowledge about the airplane, spent most of his 16 years with the Canberras in the Far East and SEA, later moving to Martin's missile program. Deployed with the B-57G to Thailand, he is shown here checking the Heading Reference System while MSgt Humphries assists.

The 500lb laser guided Mk 2 Smart Bomb was an effective weapon for the B-57G. This bomb would follow the laser beam to the target with a 0-15ft accuracy. Four were carried on the wing station. (B Knowles)

The only operational loss of the B-57G occurred on the night of 12 December 1970, deep in southern Laos. The crew of which squadron commander Paul Pitt was the pilot, and squadron sensor operator Lt Col Ed Buschette was the back-seater, safely ejected and were recovered by helicopter the next morning. All believed they had been hit by anti-aircraft fire. When a Cessna O-2A FAC airplane failed to return from the same operating area, it was concluded that both aircraft collided in the darkness.

As American forces were being withdrawn one unit at a time from SEA, most if not all the B-57Gs were flown to Clark on 12 April 1972. A month later they were ferried to the 190th Tactical Bombardment Group of the Kansas Air National Guard at Topeka. (The 13th Bomb Squadron was no longer manned or equipped but the squadron remained on the active list until 30 September 1973.) After two years of service with the ANG, the 'Gs' were retired to the reclamation depot in 1974, and after a time were scrapped.

The modifications of these B-57s as a test program was an expensive project and proved to be hard to maintain, thus the true worth of the effort may seem questionable. The effectiveness of the well publicized AC-130 with its battery of gatling guns may well

Another test aircraft in the TROPIC MOON program from which the B-57G evolved was this aircraft with the gatling gun turret installed in the belly. Sighting was tied to the sensory system that was pod-mounted on the wing pylons. These tests took place at Eglin AFB, Florida. (R. Dorr)

A number of weapon systems were tried on B-57s. One called for 52 M60C guns at varying downward angles for use as an antipersonnel and suppression weapon. Others were SUU-7; a pod under the wing for dispensing small bomblets. Another was the 'Hayes Dispenser' – a box attached to the inside of the bomb bay door having many tubes containing small bombs. When the door rotated open, the discharge of the respective tubes was controlled by an intervalometer. These and others were tested for the B-57, but none became operational.

This was the unofficial flying suit patch for the B-57G aircrews. The patch depicts a three-headed monster which related to the choice of three systems on the B-57G for tracking ground targets.

The TROPIC MOON project with B-57Gs proved the capability of having a self-contained, all-weather night interdiction bomber all as one unit. The project paved the way for later, more sophisticated systems based upon this concept. (Knowles)

The 13th Bomb Squadron with B-57Gs operated as a combat unit over the entire nineteen months that they were based at Ubon, Thailand, yet their mission focused on the operational testing of the night sensing equipment for tracking ground moving targets. (Knowles)

have proved a more economical weapon against truck traffic. But according to Bob Hunter, former 13th Bomb Squadron pilot, and later PACAF staff monitor for the night intruder missions in SEA, 80% of the bombs dropped by B-57Gs using laser guided bombs, hit within 15ft of the aiming point.

This was constructive evidence that the Air Force had achieved a self-contained, all-weather night interdiction bomber, able to deliver heavier ordnance than guns could provide. Perhaps more importantly, one concept that the test program was designed to validate was that a 'fast mover' using sophisticated electronic equipment could accurately deliver ordinance at night.

One point well remembered by Lt Col's Pitt and Strain before moving their squadron of B-57Gs to Thailand were the words expressed to them by one high ranking Air Force Official. With a steely-eyed stare came these indelible words; "I want you to understand that you are not going over there to see *if* this concept will work – you are going to *prove* in combat that it will work! Other programs depend upon it."

The fact that this concept was validated in combat was instrumental in obtaining support and funding for follow-on weapons systems like that which demonstrated pin-point accuracy repeatedly to television viewers with targets in Iraq flown by stealth bombers.

An advanced weapons system called PAVE GAT was installed in B-57G 906 in 1971 for test at Eglin AFB, Florida. This was a gimbal mounted, downward firing M61A1 gun turret which was slaved through the computer of the electro-optical sensor for target tracking. Note gatling gun barrel pointing downward from the bomb bay. Accuracy was good, but project was discontinued. (USAF via L. Davis)

21

Pakistan Air Force B-57s

How much effect the inactivation of the 345th Tactical Bombardment Group at Langley AFB, Virginia, had in equipping the Pakistan Air Force with B-57s is not known. It may well have been the needs of PAF to be equipped with bombers that hastened the closing of the 345th. In either event, in 1959, 25 B-57s from this inactivated Group were ferried to Mauripur Air Base, Karachi, Pakistan, under the Military Defense Assistance Plan (MDAP) and formed the 7th and 8th Bomber Squadrons of the 31st Bomber Wing. (See Appendix 8 for aircraft serials.) A USAF training team was sent to Pakistan for the initial training of PAF air crews.

To meet the agreed September 1959 delivery date, the airplanes were sent without the all-weather bombing systems promised by President Eisenhower during his visit to Karachi. When the RB-1A *Georgia Peach* Bombing System became available, three or four spare nose sections kept at the Warner Robins Air Depot in Georgia, were modified with the new equipment. (Two RB-57A, two B-57B and two B-57C/E nose sections forward of the canted pressurized bulkhead were on hand as spares.) This modification required a longer, more pointed shape to their nose. These were transported to Karachi and replaced the standard nose section at the point of separation behind the cockpit. The removed nose sections were returned to the U.S. to likewise be modified and the cycle was repeated until the PAF bomber force had an all-weather capability sometime after 1963. Wing racks were modified on some aircraft to carry four F-86 type drop tanks to provide sufficient range for bombing attacks on Calcutta.

With the warlike tensions that prevailed in the Middle East, and the seemingly rapid disbandment of the USAF's tactical bomb groups and wings, it appeared likely that if any American built B-57s were to see combat, they would be in the hands of the Pakistani crews. The inevitable war did erupt between India and Pakistan in September 1965, which involved these B-57s, but this was seven months after the first bombs carried by USAF B-57s were unleashed on an enemy in Southeast Asia. The conflicting reports about gains and losses in the 23-day air war by India and Pakistan are so controversial, that these claims will not be perpetuated here. It is interesting to note however, that the exchange of sorties with bombers

The counter bomber force to the Pakistan Air Force's B-57s were Indian Air Force Canberras, mostly of the Mk 8 variety shown here. Canberras began filling IAF ranks in early 1959 which prompted the PAF to be similarly equipped. (BAC)

When the 345th Bomb Group was inactivated, 25 of its B-57s were transferred to the Pakistan Air Force under the U.S. Military Assistance Program in late 1959. This Canberra, 53-3846, was one of three B-57Cs assigned with this group of airplanes. Overall camouflage is a deep green. (S. Nicolaou)

Pakistan Air Force B-57s remained in service for several years after outside aid was suspended. This was achieved by taking parts from remaining aircraft. Engine replacements became the major problem. No firm termination date for the Canberras is known. Shown here are 53-3952 and 946. (R. Walker)

were by Canberras from both sides; Martin B-57s from the PAF, and English Electric Canberras of several models with the IAF.

With the outbreak of war, all U.S. aid to the PAF was abruptly terminated. Replacement parts not already in stock were either locally manufactured or old ones repaired to keep the B-57s in a readiness condition. By 1967 the greatest impact on the B-57s was the shortage of replacement J65 engines. In September 1970, as an effort to balance power again in the troubled Middle East, the Pakistani Government was offered an option to purchase as a 'one-time exception' a limited amount of U.S. military equipment. This offer included seven additional B-57s, but this was declined. The balancing effect was short lasting, for in December 1971, the '14-Day War' erupted, which brought the PAF B-57s into open attacks against Indian airfields and military installations. Again the results were conflicting, but in both Indo-Pakistani wars, the losses of aircraft were significant, reducing the number of Canberras on both sides. As of May 1970, No 8 Squadron as well as No 31 Wing were deactivated. All aircraft were then allotted to No 7 Squadron, of which the last reliable report showed 11 B-57s were assigned. The eventual fate of the B-57s is unknown, presumably a few being saved as relics and the rest went as scrap.

This flight of three B-57Bs exchanged blow for blow against the Indian Air Force's British made Canberras in the India Pakistan War with air operations that began in September 1965. Modifications followed which elongated the nose for an all-weather bombing system. Insignia is medium green and white. (R. Walker)

22

THE LONG WING RB-57D

When 'Teddy' Petter first designed the stubby-winged Canberra, he could never have envisioned to what extremes these wings would be stretched by American designers. While the B-57 tactical bombers were beginning to take shape on the production line, word leaked out that a B-57D swept wing strategic bomber was in the making for the Strategic Air Command. There did evolve a B-57D for SAC, but it was a reconnaissance airplane sporting a straight wing, stretching 106ft, nearly twice the standard model's 64ft span.

Of the 20 RB-57Ds built, there were four distinct model Groups (see Appendix 4). Differences were slight, but seemingly all were designed for specific parts of the world for a given mission. In addition to the oversized wing, enlarged nacelles housed 10,000lb thrust J57 engines. Bomb bays were skinned over to reduce the weight of the standard bomb bay door. Wing spoilers augmented a stubby aileron, and wing flaps and speed boards were eliminated as a further weight saving measure. Despite its large wing and bigger engines, its empty weight was comparable to that of the B-57B. Honeycomb sections formed the wing in which all the fuel was carried in the wet wing and leading edge tanks. The often waxed skin was literally glued to the structure and its greatest hazard was de-icing fluid which might loosen the glue, and damage caused by the dropping of mechanic's pliers and screw drivers on the thin .010 skin surface.

Upper air sampling was one of the primary missions of the RB-57D, later designated WB-57D to signify that purpose. This view shows 53-3979 turning toward a test nuclear explosion in the Pacific to use its wing mounted air samplers at or near the cloud. (D. Bell)

Daylight high altitude reconnaissance, a mission not envisioned for the original Canberra, was also added to the B-57's versatile repertoire of assignments. This publicity picture of an RB-57D above, shows the vast wing design change required in transforming the basic RB-57A also shown here. Note the covered rear canopy indicating that this is a single seat type. Twenty were completed, of which six carried two man crews. An overpowered glider, the RB-57D was unarmed and in the rarified atmosphere above 50,000ft it could not be intercepted by the MiG-15 type of fighter, the primary threat at that time. (Martin)

Black smoke trails from the RB-57D's twin J57 engines as it forms a spectacular climb curve after take-off. In about 15 minutes the 'D' could reach 50,000ft. Take-off roll was less than 2,000ft, leaving the runway at 106kts. Best initial climb angle was about 25 deg. (J. Andrews)

While I was at the Martin Plant to take deliveries of B-57Bs, I had the feeling I should pretend that these strange looking Canberras on the ramp did not exist. Any comments made about the long, drooping winged birds was done in hushed tone. This was April 1956 and deliveries were being made to the newly formed 4028th Strategic Reconnaissance Squadron of the 4080th Strategic Reconnaissance Wing, SAC. As the specialists completed the factory training course for the airplane, they moved to Lockbourne AFB, Ohio, where most thought they were going to remain. This was not to be the case, however, for as soon as the organization was formed under Colonel Gerald W. Johnson, they relocated to Turner AFB, Georgia where the last of the SAC fighter squadrons were just been deactivated. Many of the personnel from the 31st and 508th Fighters Squadrons were absorbed into the 4028th SRS and the later to be formed sister squadron, the 4025th SRS.

At the early receiving end for RB-57Ds at Turner AFB was Stan Bunicky, a recent graduate of the B-57 transition school at Randolph, but these 'Ds' were not like any of the Canberras he had flown. However, it became his task to train the newly assigned pilots as they joined the unit. At Turner, the all silver aircraft were soon given a black and gray paint scheme unique to the 'D' model. A team of painters from Warner-Robins AFB, the depot for the B-57 aircraft, were put to the tedious task of painting. When finished, the aircraft looked like they were kin to the skunk family rather than being creatures of flight.

This was not the first unique project the men of the "Black Knights" were to encounter. Hail was familiar in that part of the country during the rainy season and hail would put the new bird out of commission in a hurry. The honeycomb used in the construction of the wing panels would not endure a hail storm. The wings had to be covered, but the problem of procuring covers that would not corrode the metal was not easily found but was eventually resolved.

Some of the other problems that plagued the new organization were; autopilot failures, engine bleed pressures, a ruptured wing tank, nepho equipment malfunctions, lack of trained people compared to the work loads, and one case of the bends. In all they totaled to an average of over 5500 man-hours per month overtime from September 1956 to the end of that year.

By the time the squadron was declared operational after a mere 120 days following receipt of the first aircraft, six of the airplanes

Martin RB-57D 'Zero' Canberra (Groups A and B)

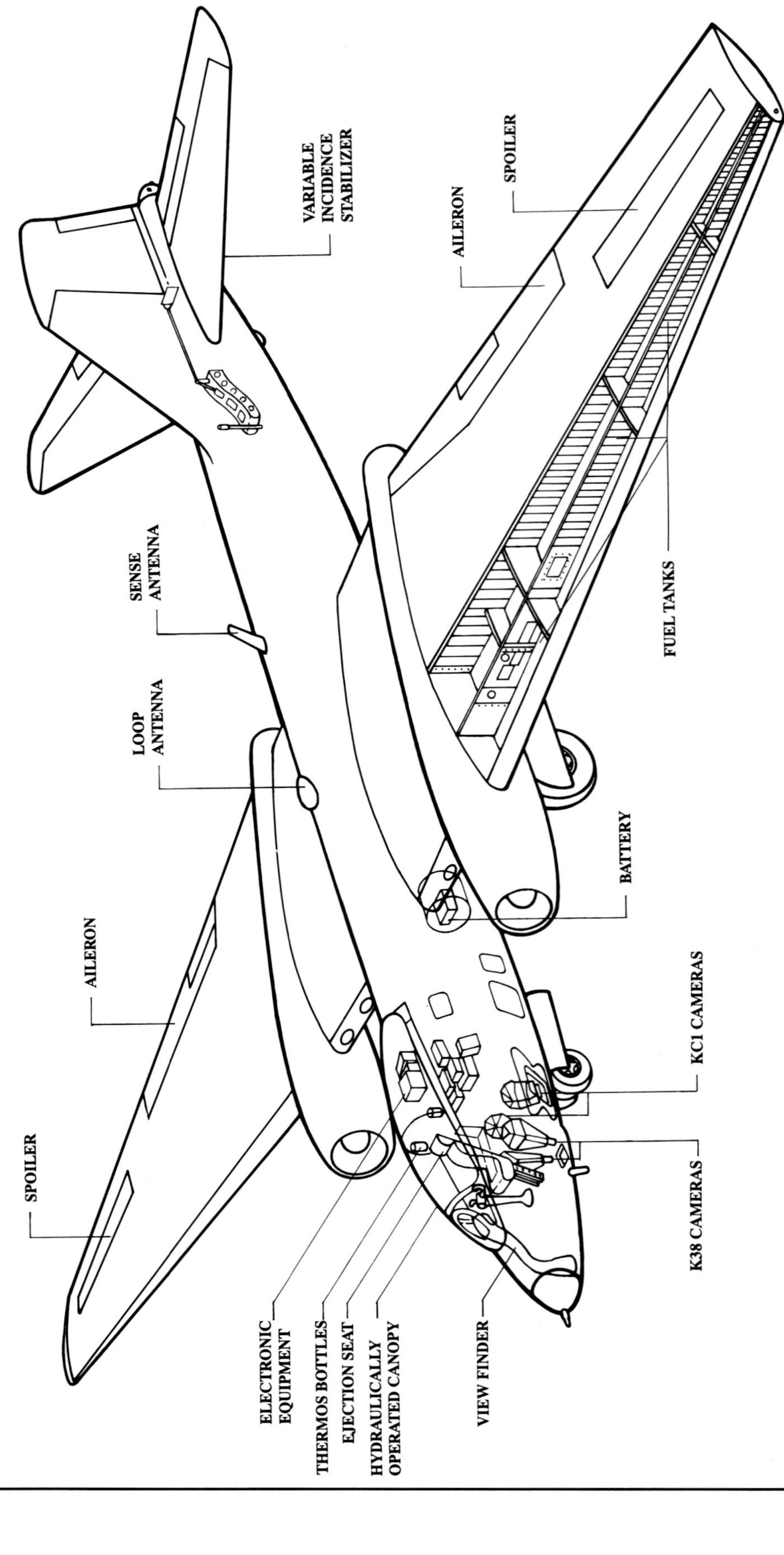

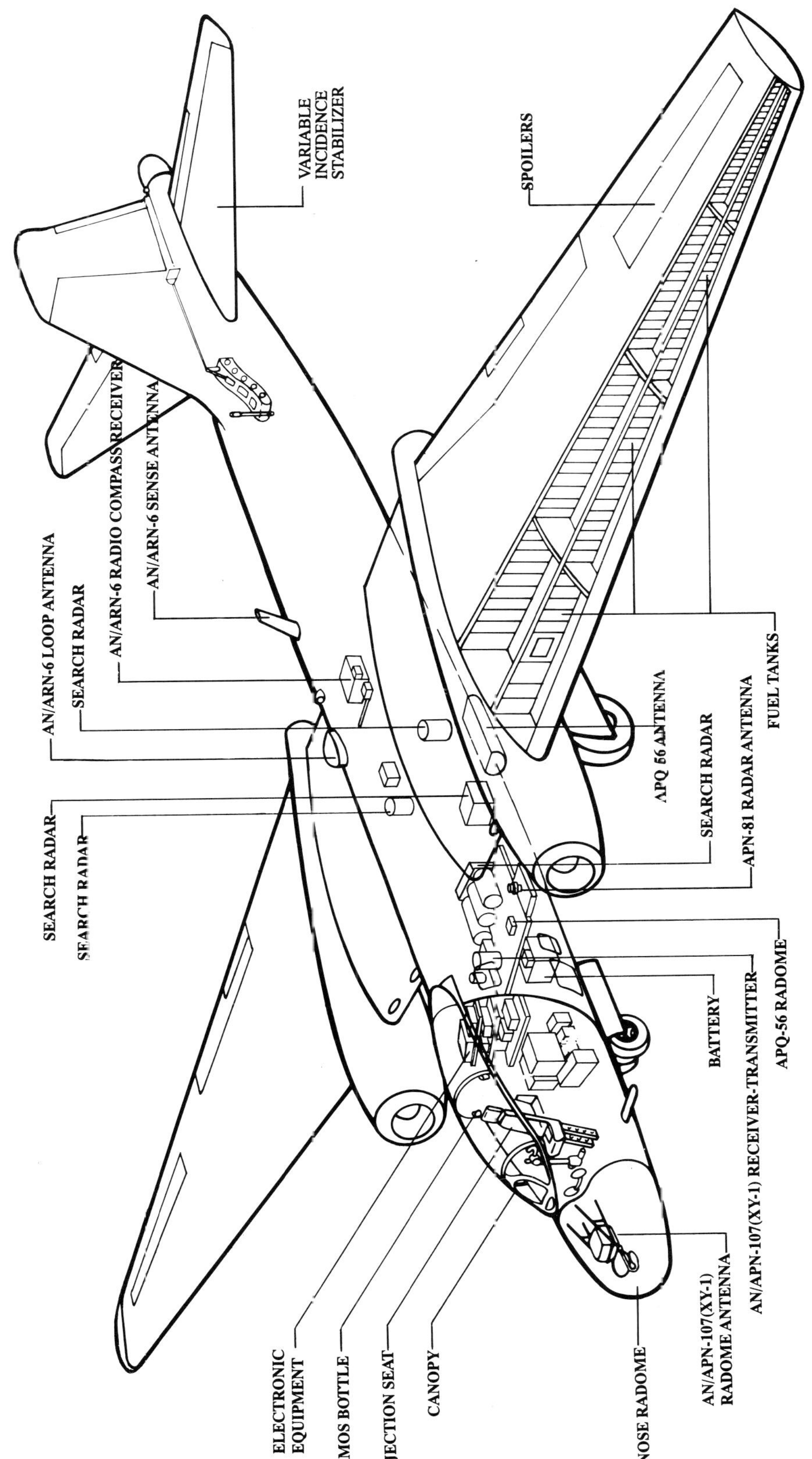

Martin RB-57D-1 Canberra (Group D)

Martin RB-57D-2 Canberra (Group C)

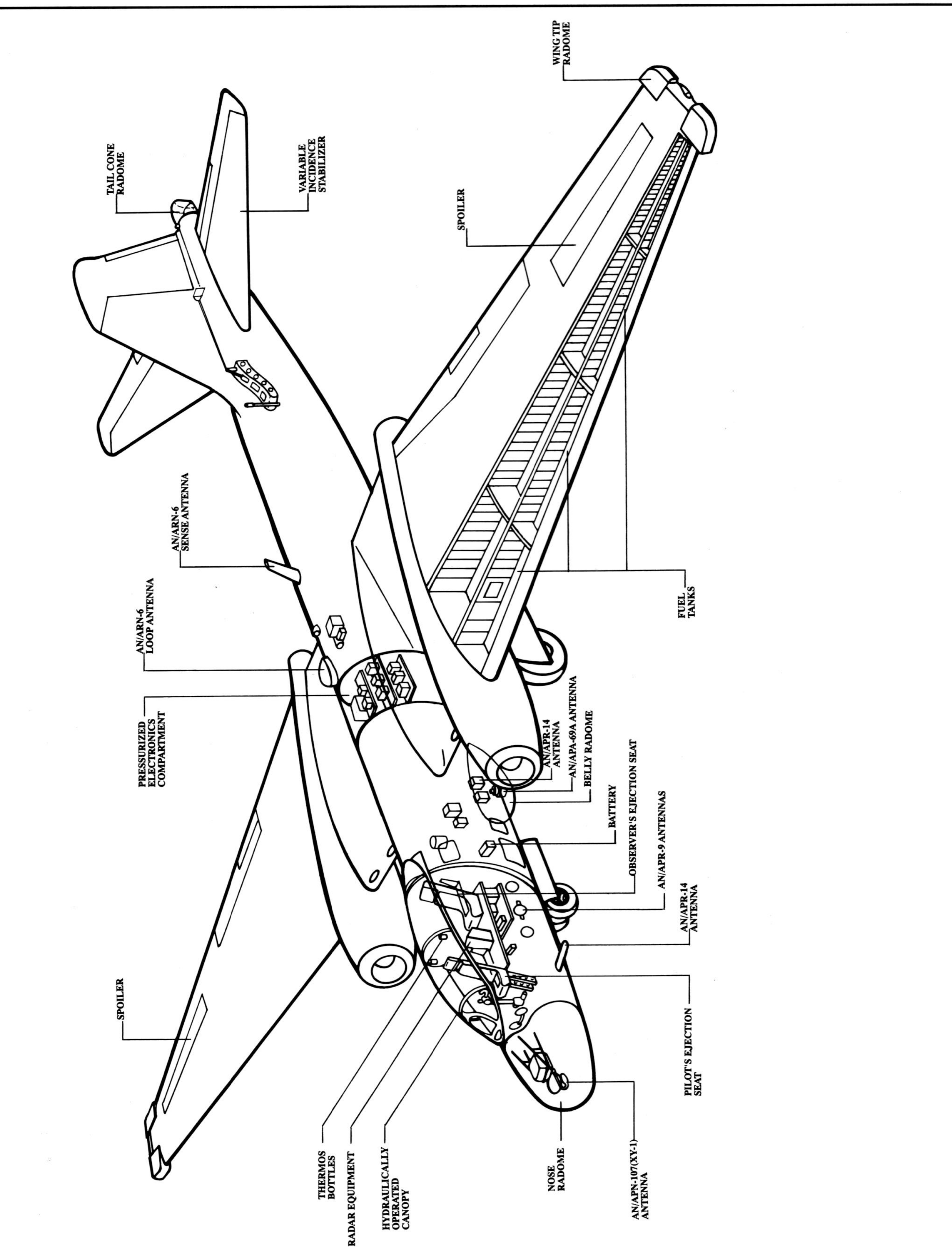

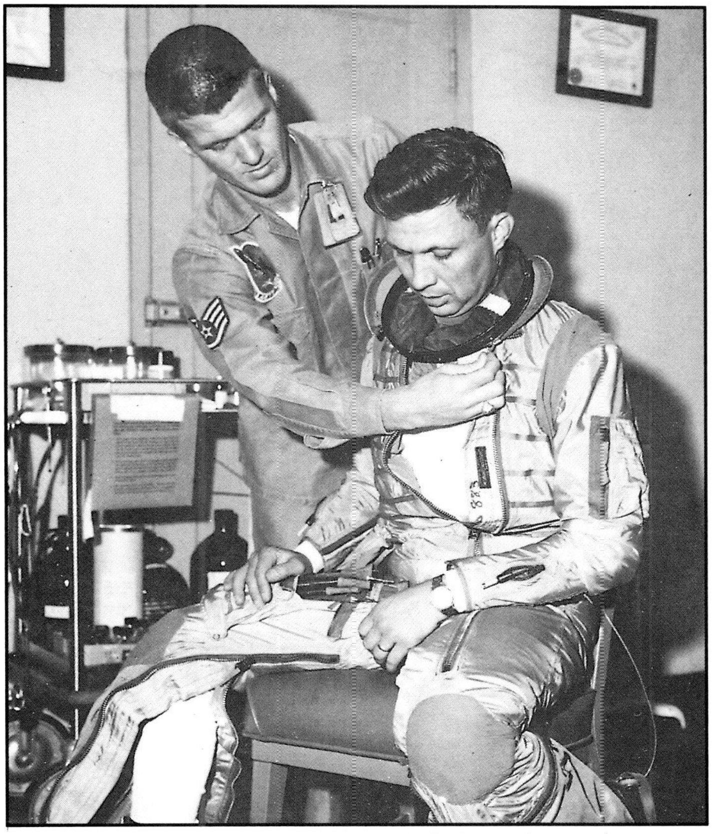

Pressure suits required for all high altitude flights in the RB-57Ds were always a task to get into. Capt Homer Caldwell of the 4677th DSES in 1962, gets into a capstan partial pressure suit variety that was later replaced by CFU-4P experimental suits. All were commonly called 'Get-me-downs' which related to anxiety of crews to land and get out of their discomfort. (E. Bruch)

were deployed to Yokota AB, Japan. This deployment began on 11 September 1956 for what was called Project *Sea Lion*. These were of the Group A airplanes of the photo reconnaissance versions with only a pilot, and without in-flight refuelling. These SAC airplanes supplemented the Yokota based 6021st RB-57A-1s with their higher thrust J65 'Heart Throb' engines for their strategic reconnaissance mission in the sensitive areas of the Far East. While en route from Turner to Japan, one aircraft landed gear-up at McClellan AFB, California because of mechanical problems. This dampened the spirits of the men somewhat but they continued to forge ahead with their new birds.

Beginning in early 1957, the Wing and its squadrons moved to Laughlin AFB, Del Rio, Texas. With the men and airplanes scattered to many parts of the world, just how are families, dogs, cats, birds, and tank fish moved hundreds of miles, and still maintain some resemblance of an organization? For those residing at Turner, they made the move to Laughlin, then went TDY to relieve the group overseas so that they too could return to Turner and move their families to Texas. It was a difficult time for all concerned.

In close order, the Yokota based RB-57Ds left the scene in late 1957, only to have about three of them reappear in late 1958 on Taiwan, at Taoyuan AB, near Taipei. Carrying Chinese markings, they made frequent reconnaissance flights over the mainland during this period of the Taiwan Straits Crisis. At least one was shot down when making a premature descent back to Taiwan, while two are known to have returned to the U.S. when the program ended around 1963. At the onset, Ellis Bruch and Bill Bunting of the 3rd BW took two B-57Cs from Japan to Taiwan and trained two Chinese pilots to fly the Canberra. This checkout program code-named *Diamond Lil* ended abruptly at just under 30 days when the two sides began air engagements. According to Bruch, the pilots were top-notch and may well have managed their own training for the 'D' from that point on.

After an absence in the Far East of about one year following assignment in Japan, three 'Group A' RB-57Ds took up station on Taiwan in 1959 prompted by the Taiwan Straits Crisis. This time they were clad in Nationalist Chinese markings and operated at Taoyuan AB, Taiwan, carrying out high-altitude reconnaissance flights over the mainland. The author took this picture of 981 when taxying by in another B-57 on this heavily guarded base.

The main drawback of the RB-57D was wing failure like this one of two that literally broke off after landing. A third occurred at altitude, grounding the fleet a second time. Some were structurally beefed-up and remained in service until 1970. (D. Anderton)

On 14 April 1958, six RB-57Ds, crews, and support personnel and equipment were in place at Eniwetok Atoll (Fred Island), Marshall Islands for Operation *Hard Tack*. They were to participate in the atomic bomb tests conducted in that remote part of the world. The last of this Det. 7 of the 4080th did not return from these tests to Laughlin until 11 August of that year.

On the other side of the world, four of the Group B aircraft with in-flight refuelling capability, were sent from the States to make occasional visits to Rhine Main, Germany. Their missions were long and for this reason the airplanes were equipped with an autopilot, and rudder pedals that could fold, giving the pilot room to stretch his legs. Two more 'long wings' joined the unit, one being the only RB-57D-1, 963, featuring a large nose radome and lengthy sausage-like radomes faired into the center of the fuselage under the wing roots. Its special high resolution side-looking radar capability was obviously designed exclusively to penetrate the Iron Curtain. On 9 June 1959, six of these airplanes were permanently assigned to Rhine Main as part of Project *Big Safari*, forming the 7407th Support Squadron, and assigned directly to USAFE Operations. All RB-57D operations were closely guarded, and they received only a trace of publicity in their early operations. They too were termed the 'Spy in the Sky' along with the U-2.

The long wing model began to lose favor when wing failures occurred and SAC placed several in storage by early 1959. Two outer wing panels literally broke off after landings at Del Rio and Kirtland AFB. Their 500-hour mark for the designed airframe life was being surpassed by some, but a few were retained to continue with the mission out of Rhine Main until 1964. In the meantime, the Air Defense Command saw the potential of this high altitude airplane as a 'target' for the increasing capabilities of fighter-interceptors. Modifications were made to strengthen the wings of the grounded 'Ds', and Electronic Counter-Measures equipment was installed. The 4677th Defense Systems Evaluation Squadron of ADC at Hill AFB, Utah, received 12 of the RB-57Ds over a period of time which joined the 30 standard wing models of that unit. These aircraft filled the 'target' training mission well, and also mixed in a small amount of high altitude photography. Another use found for ADC RB-57Ds was to supplement SAC assigned 'Ds' for continue-on cloud sampling in the vicinity of nuclear bomb tests as with previously described Operation *Hard Tack*. (As early as 1954, RB-57As were used for Operation *Castle* in Nevada for this sampling, followed in 1956 by Operation *Red Wing* at Eniwetok using B-57Bs.) All models of the B-57 proved very useful for this type of work, and the feature liked most about the 'D' was its ability to carry a good-sized payload of sampling sensors higher than any other current aircraft.

After a three-year period during which nuclear testing had been discontinued, plans for tests resumed. This detonation test called Project *Dominic*, took place in 1962 and, unknown at the time, was the last above ground nuclear shot by the U.S. The 4677th DSES at

The long wing span is emphasized in this picture of a 4677th DSES 'target' RB-57D. Hose to port engine extends from ground-power unit used for air starting the J57 engine. In pressure suit in front of engine is Jim Scanlon preparing to depart from Hill AFB, Utah. (J. Scanlon)

On Christmas Island in the Pacific, B-57s gather in 1962 for the nuclear detonation test called PROJECT DOMINIC. Eight black RB-57Ds shown in this picture were used for reaching very high altitudes for cloud sampling in the vicinity of the test area. White B-57Bs tested the radiation effect of the blast as if they were in the escape location after delivering such a bomb. (W. Boyne)

WB-57Bs served along with WB-57Ds in the high altitude air sampling missions after nuclear detonations in PROJECT DOMINIC in 1962. Decontamination teams wash down 504 after its cloud penetration sampling.

The Air Defense Command obtained a number of RB-57Ds and converted them to EB-57Ds as intercept targets for fighter training purposes. This EB-57D-2 now stripped of intelligence gathering radar, retained its radome nose configuration.

Cockpit of this EB-57D-2, 968, is very similar to the standard wing models. On the right console are the auto pilot controls not found in earlier models. Control wheel had a comfortable, sturdy look and feel on all models.

Hill AFB provided five 'Ds' along with crews. These were joined by the one 'D' from Wright Patterson AFB (973) assigned there for flight test projects. (WADC received three RB-57Ds from the 4080th SRW beginning 16 June 1959.) Six of the standard wing B-57s were also at Christmas Island for this mission. The 1211th Test Squadron (Sampling), Air Weather Service, had operational control over the loaned aircraft and in time acquired permanently assigned WB-57s for the sampling program. This Kirtland AFB, New Mexico, based unit eventually became the 58th Weather Reconnaissance Squadron.

Structural problems in 1963 put most of the 'Ds' back in storage for a second time when all were grounded after 973 lost its

This Martin EB-57D sits at rest on the ramp of the 4677th DSES at Hill Air Force Base, Utah in the early-1960s. It was at this time that day-glo orange paint became standard high visibility coloring applied to wing tips, nose and aft fuselage.

RB-57Ds were used as a platform in a wide range of special high altitude testing of various equipment. This aircraft, 53-3973, shown as it taxies at Wright-Patterson AFB. On February 19, 1964, it had the misfortune of having both wings separate from the fuselage at 50,000ft. The pilot escaped, and although the fuselage landed in a school yard, no one was injured. (Picciani)

wing at 50,000ft. Years passed, and ADC still had a requirement for a high altitude 'target' aircraft, seemingly only to be satisfied by the 'Ds' performance. Martin agreed to modify the wings of the 'D' and guaranteed an additional 3,000 flying hours. Only the 'D-zero's' and D-2s could be made airworthy for the flight to Martin for modification. When completed, it was the third time these airplanes were brought into operational service. Updating of their ECM (Electronic Counter Measures) gear brought about a redesignation to EB-57D. This 'long wing' model served other test and evaluation missions throughout the Air Force, but by mid-1970, the last were again placed in storage. In time all were scrapped except for 982 which is now at the Pima Air Museum, Tucson, Arizona.

Piloting the RB-57D was a unique experience in itself when compared with other aircraft. Only half power was used for take-off due to the excessive thrust of the larger engines. Should one engine fail at low airspeed and at full power, corrective reduction in power of the other engine could not be done quickly enough to maintain control of the aircraft. After a ground roll of less than 2,000ft, lift-off at 110kts resembled that of a sailplane, then transitioning to a high nose attitude, it was out of sight in a matter of moments. Everything stopped on the field to watch these spectacular take-offs. Initial climb would go to 57,000ft as the optimum altitude for level off. For the best range, cruise-climb would continue until maximum altitude of about 65,000ft would be reached. This capability was slightly less than that of the lighter U-2. Despite the much larger engines on the 'D', fuel consumption at these heights was very low. Climbing in this manner to these altitudes, approximately seven hours of flying could be obtained with the total fuel capacity of about 214 U.S. gallons less than that of the standard B-57B that could fly for about 4.5 hours.

The airplane had a maximum indicated airspeed limitation of 190kts that was further reduced to 180kts IAS when the outer wing auxiliary wing tanks became empty. Although this reduced airspeed seems exceptionally low, it worked out at 420kts TAS plus at its operating altitude.

Landing the airplane was quite different from any other because of its low sink rate. Idle power at traffic pattern altitude was still 60% far more than actually needed. Even with spoilers out and gear down (there were no speed brakes or landing flaps), to dive the airplane on the glide path did little more than increase the airspeed. By raising the nose to produce a slight stall was the best method to hold the airspeed down and to let the airplane settle back to the glide path.

According to Jim Scanlon who flew the 'D' and other models of the B-57s for many years, 'If ever an airplane needed an anchor and a long string to get it down on the runway, the EB-57D was that airplane.'

This picture was taken as RB-57D-1, 53-3963, moved down the taxiway at Rhein Main Air Base, Germany. It was assigned there with the 7407th Support Squadron. This one-of-a-kind model was equipped with SLAR (Side Looking Airborne Radar) that can be seen where the trailing edge of the wing joins the fuselage. (R. Koch via D. Menard)

23

GENERAL DYNAMICS 'F'

As the service life of the RB-57D was running out, so would end a vital intelligence gathering mission flown by the 7407th Support Squadron at Rhine Main. The arduous task of maintaining these Canberras for the Air Force was the responsibility of the Fort Worth Division of General Dynamics. Making frequent repairs and updating the 'Ds' internal equipment gave General Dynamics considerable experience with the ailing airplanes and they devised a number of ways in which improvements could be made for continuing the mission.

It was not until several new requirements for a 'super D' came about that the Air Force really became interested in the GD/FW proposals. In the early 1960s world wide attention was being focused on the Middle East, and U.S. aid was directed toward Pakistan. From this generated Project *Peewee*, a highly classified re-

The enormous wing of the RB-57F, while awkward on the ground, gave a sense of grace in the air. NASA made use of the 'F' from their inception to becoming the last to fly the American Canberra. Dissimilar air samplers are attached to the outer wing panel pods.

From this angle, the RB-57F, later redesignated WB-57F, looks to be only wing and engines – which proportionately it was. Engines are P&W TF33 turbofans furnishing 16,000lb each. (L. Mongeon)

connaissance gathering mission calling for a high altitude reconnaissance airplane. In addition, nuclear tests were still being conducted, and Joint Task Force 8, an agency responsible for nuclear testing and gathering of results, also had a requirement for a high altitude aircraft.

On 27 March 1962, the Air Force awarded a study contract to GD/FW for the RB-57F concept. The object of the study was to obtain, from the basic RB-57D, improvements in performance, maintainability, and an increase in service life. What evolved was a design having even greater wing span than the already exaggerated 'D', reaching 122ft and a wing area of 2,000sq ft as compared to 960sq ft on the standard model. Satisfied with the design feasibility, the Air Force awarded a contract for building two RB-57Fs to GD/FW on 2 October 1962. First metal was cut on 28 November, and seven months later on 23 June 1963, Capt E. 'Ted' Sturmthal made the first flight of an EB-57F. This was the first time that an Air Force pilot made a first flight in a new airplane. Ted was with the program from the very beginning and remained throughout the development test phase. By February 1964, the 'F' was pronounced operational and two were quickly sent to Pakistan, and shortly afterwards were said to have been operated for or by the CIA.

Production of more RB-57Fs followed. Helping to finance the project was the Atomic Energy Commission since the airplanes would be used for future atomic cloud sampling. (Oddly, this requirement never materialized.) In all, 21 were built, utilizing only the original fuselage, horizontal stabilizer, and landing gear from the scarce supply of remaining B-57s. All 14 B-57Bs selected for conversion were from the active fleet, while four RB-57Ds; two formerly with the 7407th Support Squadron, were taken from storage at Davis-Monthan AFB. Three RB-57A aircraft were converted to 'Fs' utilizing three spare B-57B nose sections (described in the chapter on Pakistan B-57s). Production was completed in March 1967. (See Appendix 5 for airframes converted.) Power for the 'F' was more than doubled by the Pratt & Whitney TF33-P-11A turbo-

General Dynamics-built RB-57F was the ultimate in expanding the design of the basic Canberra airframe. With new wings that span 122ft, it more than doubled the original wing area. This was actually a four-engine airplane having two detachable pod mounted J60 engines under the wing. Wing tips are formed from fiberglass. The Air Weather Service used these airplanes for high altitude air sampling as well as photography and other high altitude research projects. (NASA)

fan engine, furnishing 16,000lb of thrust each. Take-off power was limited to 68% RPM. This gave safe single engine airspeed of 120kts for that power setting with still a relatively short ground roll. The F' actually became a four-engine airplane with the addition of two detachable Pratt & Whitney J60-P-9 turbojets. These produced about 3,000lb st at sea level and were under each wing outboard of the main engines. The J60s were normally in place, but they were removed for maximum range missions. Not having starters, they were air-started after take-off (except for maintenance functions) after windmilling to 12%. They remained at idle until 32,000ft where throttle control started becoming effective. Full throttle control was available above 42,000ft. They added approximately 2,500ft more altitude, which was comparable to the U-2 of that era.

In the design of the 'F' model, GD not only made use of their 'D' experience, but also the knowledge acquired while building the B-58 Hustler's wing. This same light-weight honeycomb sandwich skin was incorporated into the 'F', using the earlier tools, fabrication and assembly techniques, giving a saving in production costs. Instead of relying on the one piece wing spars as Martin had with the 'D' in expanding the original wing, the new wing for the 'F' consisted of a three spar design further strengthened with the honeycomb skin structure. The replacement fin and rudder which nearly doubled the original surface area, were constructed with this same technique. In addition to improving single engine control, this increase was primarily to dampen the dutch roll cycle at altitude caused by increased wing area which had been objectionable in the RB-57D.

Airframes that followed the first two RB-57Fs were primarily Air Weather Service sampler versions. Installed equipment included four gas samplers, two air samplers beyond the mid-wing point with paper filter elements, three radiation meters, and voice and data recorders. For other type missions, a F-415P vertical panoramic camera was installed.

Special personal equipment – far from being comfortable – was essential on all high altitude flights in the WB-57F. Suited up, Lee Mongeon happily looks forward to seven or eight hours in the Dash 2 pressure suit and WB-57F of the 58th WRS, Kirtland AFB, NM, in 1966. (L. Mongeon)

The 58th Weather Reconnaissance Squadron at Kirtland AFB, Albuquerque, New Mexico, situated near the Los Alamos Scientific Laboratory (LASL), one of the AEC laboratories they serviced, was the main user of these aircraft for the 10 years they were active. They received their first on 17 June 1964. These RB-57Fs, later redesignated WB-57Fs, served in many parts of the world including: Germany, Japan, Australia, Spain, Argentina, Johnston Atoll, Panama and Alaska, naming only the major overseas locations. During periods of known or suspected atmospheric testing of nuclear devices anywhere throughout the world, the WB-57Fs were on hand for nuclear debris sample collecting in the upper atmosphere. These samples when analyzed have provided valuable diagnostic data from foreign and U.S. nuclear detonations. The scientific direction for these aircraft was provided primarily by Dr. Paul Guthals of LASL. Guthals managed the scientific supervision

The long wing of the WB-57F often overshadows the many other details shown in this close-up view. Nose was lengthened 40in to house radar, cameras and air sampling equipment. Canopy curtains protect crew from sun radiation in rarified air. This WB-57F, 294 was formerly 935, a 'B' model assigned to the 8th Bomb Squadron, Japan. (N. Taylor)

Early warning that the nose wheel would not lock due to mechanical failure of this WB-57F, allowed time to foam the runway to reduce aircraft damage. Cost of repair was under $10 for fiberglass patch to nose. A high resolution camera (not considered part of the aircraft) took a terrible beating however. (L. Mongeon)

The mammoth wing area of the WB-57F is boldly apparent in this plan view picture of NASA 928, AF serial 63-13298. Note the J60 engine pods protruding outboard of the main engine nacelles. (NASA via S. Nicolaou)

Although few in number WB-57Fs could be expected to be seen in any part of the world depending upon mission requirements. Beginning in mid-1960s, three WB-57Fs, including this aircraft, 63-13503, were assigned to the 6091st TRS at Yokota AB, Japan

of the 58th WRS sampling activities both on the ground and, with others, as an airborne Scientific Mission Director.

The 500 series of the 'F' were designed for specialized reconnaissance duties, two of which; 501 and 503, were assigned with the 6091st TRS at Yokota AB, Japan, beginning 16 May 1965, while 500 and 502 went to the 7407th Support Squadron at Rhine Main. It was from this squadron that the first loss occurred. Aircraft 287 crashed on 14 December 1965 in the Black Sea at a time when tensions ran high in that part of the world. Russian and American ships vied with one another in recovering the parts of this new American spy plane. Les Lackey and his crewmate Bob Yates were listed as MIA for six months after the crash and then declared dead. The incident was kept very quiet and the true cause – SAMs, structural, or crew factors – has not openly been revealed.

Two other 'Fs' were also lost. In November 1966, 297 piloted by Bob Bartlett crashed into Sandia Crest near Albuquerque. With one engine out, low on fuel, and in the clouds at night, Bob let down too early and hit the crest at the 10,600ft point just a hundred or so feet below the summit. The final loss occurred in mid-1972, also near Albuquerque, with Lt Col McGravey as pilot. The air-

Nose access compartment of WB-57F contains air sampling equipment. At right is Paul Guthals, highly regarded for his scientific supervision with Los Alamos Scientific Laboratory for air sampling techniques used throughout the many years that B-57s were used to support this mission. (LASL)

Simplified loading of specialized equipment is facilitated by a removable pallet attached to the fuselage cavity in the same manner that bomb bay doors are handled. Joe Harwell with Lockheed for sensor maintenance support, mans one of four detachable hoists used for removing and attaching the 4,200lb pallet. (D. Anderton)

RB57F 13501 NASA PALLET

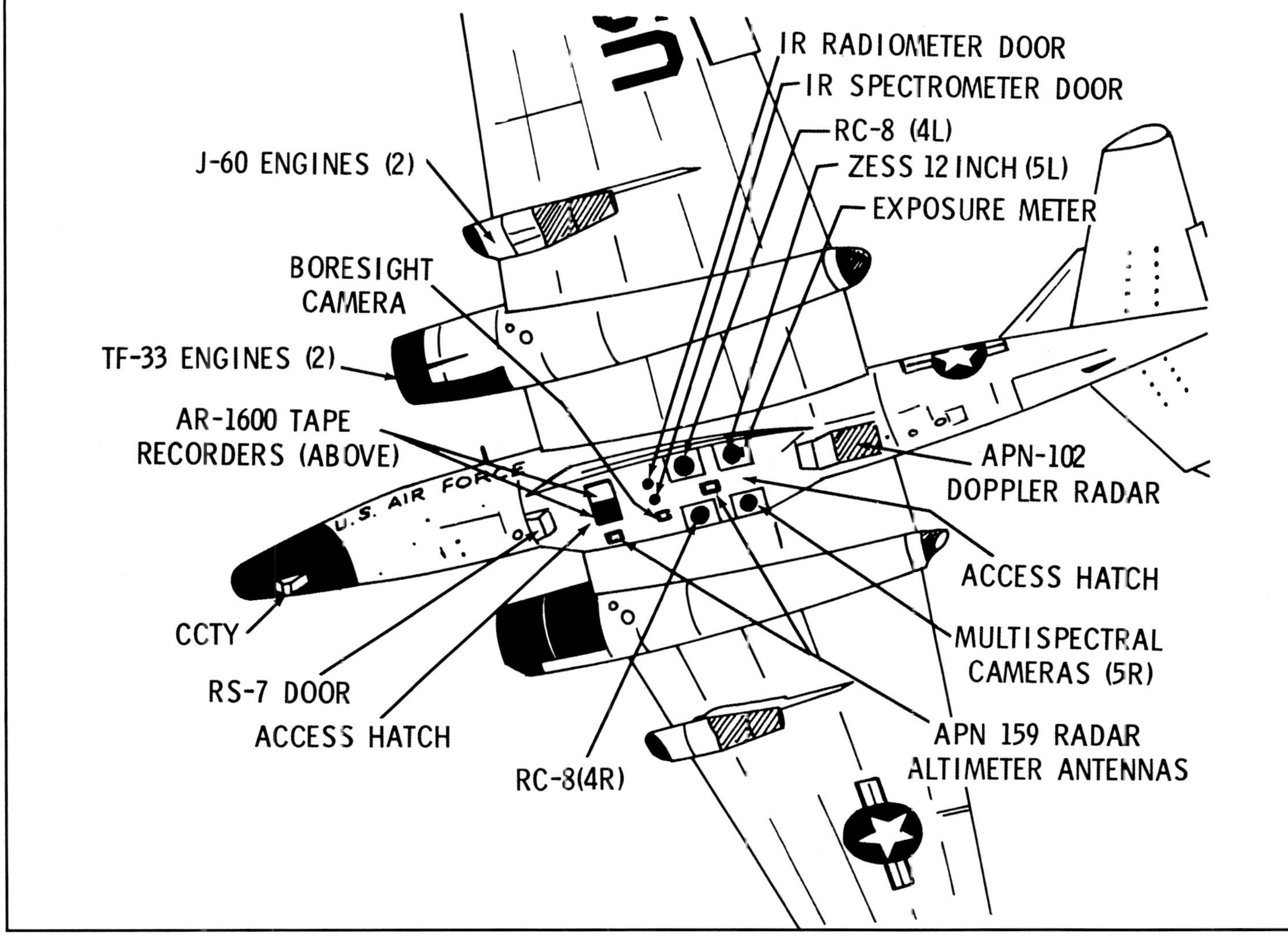

plane inadvertently entered a Mach tuck followed by wing flutter and came apart around 50,000ft. One 'F' was badly damaged and nearly destroyed in a hangar at Karachi as a result of a bomb dropped by an Indian AF Canberra in December 1971.

Unlike the early 'Ds' with normal corrosion problems, stress corrosion cracks began appearing on the wing spars and ribs of the 'F' after a few short years of service. Some of these aircraft were sent to GD/FW and Hayes Birmingham for repairs. The basic problem was the 7,000 series aluminum alloy used in the wing structure developed stress cracks through time alone. The repairs changed wing spars and ribs to 3,000 series aluminum. Due to the excessive cost to repair all the aircraft, coupled with programmatic changes, nine of the unmodified fleet were placed in storage at Davis-Monthan in the spring of 1972. For the rest of the fleet, the end of their flying days was near. The 58th WRS, being the last in the Air Force using the WB-57F was deactivated by 1 July 1974 after placing their Canberras into inviolate storage with the others at Davis-Monthan.

This did not end the era of the long wing 'F' however, for one or two continue to fly at this writing. As early as September 1968, NASA had contracted with the Air Force to operate a RB-57F for them in connection with their ERTS (Earth Resources Technology Satellite) program. The airplane placed on loan was 63-13501 with USAF furnished pilots and maintenance. NASA provided the second station special equipment operators and maintained their sensor systems. Modifications needed to this reconnaissance airplane were basically installing bracketry so that a data gathering sensor pallet could be raised and attached in the center fuselage area. This pallet changed the bottom of the center fuselage to a flat surface to facilitate the mounting of cameras and electronic scanners. The pallet was faired in so well that there was no increase in drag. The cost to NASA for this mod was $3.8 million, and the airplane went into service in July 1969.

By January 1972, the Air Force decided it was costing too much money to share the expense with NASA in operating this aircraft, so transferred it to them. This became NASA 925, which was their Earth Survey Aircraft (ESA) No 3. NASA continued to operate this airplane for the next eleven years until 1982 when it was retired. It was not to be forgotten, however, for after being delivered to Davis-Monthan AFB and AMARC at Tucson, Arizona, it was placed on

At the time of this book printing, EB-57F, NASA 298, and possible NASA 926 are the only survivors of the B-57 family that are operational today. Their termination date is still not planned. The new paint scheme that appeared here on 298 in 1993 was for surface protection because the original metal coating was nearly rubbed off from cleaning. (NASA)

loan by the Air Force to the Pima Air Museum across the road where it resides today with many other noteworthy aircraft.

As the 58th WRS was phasing down in early 1974, 63-13503 was also transferred to NASA and became NASA 926, ESA No 6. In its later years of service, and among other and varying missions, it was used as NASA's Earth Remote Sensing Platform for calibrating satellite data. When funding for this major project was no longer available, it was placed in hangar storage at Houston, and later moved to El Paso in 1991 where it is now stored. Prospects are likely that it will be transferred to the National Center for Atmospheric Research (NCAR) where it will most likely be an air sample platform and operate from Boulder, Colorado.

On loan to NASA was a third long wing 'F', 63-13298, renumbered to NASA 928. This aircraft was and has remained a sampler. The Department of Energy (formerly Atomic Energy Commission, and Energy Research and Development Administration)

Long after these WB-57F Canberras were placed in storage at Davis Monthan AFB, Arizona, they were maintained to be regenerated if needed. Parts taken from these aircraft kept others flying. After a number of years, the Air Force no longer had a need for WB-57Fs and they were scrapped.

makes use of this airplane much as they did when the USAF operated the WB-57Fs. Like the others, this too was eventually transferred to NASA. In recent years, this has been the only B-57 of any type still flying, and makes for a very small and unique group of qualified B-57 pilots, ranging from a mere two to four.

The current project pilot is Steve Feaster who has been with the program since 1982. He reports that in recent years, NASA 928 has logged approximately 300 hours per year which is quite high for a specialty aircraft of this type. The aircraft's most recent involvement has been Project *Air Stream*, using DOE air sampling sensors. Current plans call for the airplane to continue to record earth distortion that is derived from simulated large explosions, coupled with related air sampling. Tracing volcanic debris dispersion is another important mission performed by this airplane.

When NASA first began using WB-57Fs in 1968, Air Force pilots were assigned with NASA as a joint effort to fly these missions. Serving as project pilot at that time was Major Robert Danielson from the 58th WRS. When asked to describe how a mission was flown for NASA in these Canberras, Bob gave this account:

> 'A typical flight take-off would be from the operating base closest to the area of interest. We would photograph the area and/or collect electronic data, and return to the operating location. Most every flight was data productive or at least planned that way. When deploying the aircraft to an operating location, a data acquisition site was planned into the flight so as to be as effective as possible.
>
> 'For a high priority requirement, we would fly from home base and gather in-flight data at a distant site, land at one of the operating locations, refuel, and a second crew would fly the bird home that same day. This way we could have the data processed that same day.
>
> 'One publicized series of missions was data collecting over the corn fields of the mid-west for a project called "Lacie" (Large Area Crop Inventory Experiment). These data were used in the control of corn leaf blight and other agricultural maladies.
>
> 'Flight crews were required to wear full-pressure suits which complicated the operation somewhat, ie, a second crew was required to be on duty to preflight the aircraft and be prepared to fly in case the primary crew did not pass the brief physical exam requirements before each high altitude flight. Also, a pressure suited crew could not fly twice in one day, and a flight duration of 5hr 30min or more required a mandatory crew rest period of 30 hours. Therefore a crew could not fly on two consecutive days if this limit was reached. Under optimum conditions, the aircraft could be flown for approximately 7hr with fuel reserve. This included one hour for the climb phase 5hr 30min for data collection at altitude, and 30min for descent and landing.'

24

B-57E Cadillac

All models of the B-57 were not intended for sophisticated reconnaissance or direct encounter with a potential enemy. One such Canberra model was the B-57E, configured on Martin's production line for towing targets. When the first one of these airplanes appeared on our ramp, we admired it like the neighbor's new car. By early 1957, we were accustomed to our B-57Bs smudged and eggshell cracked black paint. The 'E', fresh from the factory, glistened with its shining brilliant orange upper surface and silver painted lower half. I liked flying the 'E' better than the earlier models, for it did have minor improvements, but in truth it was probably because they were newer and cleaner than the tactical version.

Structurally, the tow target model was a carbon copy of the B-57C dual control model, but without its combat gear. It had tow reel operating equipment installed in the back seat in place of the Shoran bombing system. Additionally, it had a full-time power rudder (allowing a reduced single engine airspeed of 135kts) and yaw damper, and the tail cone was altered slightly to accommodate the two externally mounted target canisters. Sixty-eight of these models were built by Martin on a separate contract.

Northrop designed and built the armor plated tow reels that were mounted on the bomb bay door. Why armor plate steel? At the rate a steel cable-end could whip around the drum within the thin skinned aluminum fuselage and under the volatile fuel tanks, protection was imperative. The bomb bay door would not open on this model and had to be lowered to the ground for servicing. There were four 1,500lb reels mounted on the door, having their centerlines running parallel with the fuselage. Add 1,500lb of armor coated tow cable to each, measuring 6,000ft, and you have a heavily loaded airplane, grossing around 55,000lb.

The standard bomb bay door served as mount for the target tow-reels. Tow reel operators at Johnson AB, Japan in 1958 wheel the door under the B-57E for hook-up after servicing air turbine drive motor that turns the cable drum.

With a target banner close in tow, this photograph was a popular promotional poster announcing the new Martin B-57E tow target aircraft. Following bomber model production, 68 of this version were made to support the fighter-interceptor training mission. (Martin)

Target launch method was relatively simple. The cable drum which was driven by an air turbine for in only, would draw in a short amount of cable that was threaded out through the tail grommet and attached to the forward end of a target in canister. Once the target was drawn out of the cannister into the air stream, it would unfurl and the turbine would act as a break to the cable drum until extended to about 5,000ft. Shown here is the loading operation. (USAF)

This modified B-57E of the 17th TTS, Yuma, Arizona in 1958, had tail mounted baskets to hold two Delmar frangible radar reflective targets instead of banners. These styrofoam targets required only 1/16in cable and due to their light weight and low drag, were towed as high as 40,000ft (and higher) at .74 Mach. F-86Ds, F-89s and F-94s fired 2.75in rockets, Genies, and radar guided missiles. Tyndall AWC also used this modified system on their B-57Es. (L. Mongeon)

Launching the target usually started no higher than 30,000ft so there would be sufficient 13th stage air pressure off the engines for operating the turbine on the tow reel. To launch the target, the tow reel operator seated behind the pilot, slowly advanced the system throttle which controlled an air metering valve. Enough air to move the reel a few turns brought in sufficient cable to draw the target rearward out of the canister. It was here that the skill of the operator counted – not to allow the tow reel to overspeed on the way out, and not to over correct by adding too much air as a brake to the drum's turbine and cause the target to come back to the tail grommet and break off.

When the target was fully deployed, the drag was tremendous with 5,000ft of 3/16 in cable in tow with a 9ft x 45ft banner target attached. It was a challenge to regain 200kts IAS from behind the power curve at 130kts target launch speed. I recall operating the engines at 100% for the entire first hour of each target towing flight until they could slowly be worked back to 96.5% 'max continuous' as fuel weight burned off. This was a heavy demand on the engines but they held up under the strain.

A chase aircraft, usually a T-33, would drop back occasionally to score the target for holes caused by 2.75in rockets fired by F-86Ds. When firing was completed, the cable was drawn in, bringing the target to within a few feet from the tail. Electrically operated hydraulic cutters would sever the cable and release the target. A second target could then be launched followed by a third and fourth if there was a requirement. Normally, two targets filled the

This Martin B-57E has just launched a 9ft x 45ft banner target that will trail 5,000ft on 3/16in armor coated cable before fighter firing passes begin. These 'Es' were built specifically for towing targets, yet had the capability for conversion to the combat role – and some were. Both these aircraft served in SEA, 269 with the 8th BS and 264 with 'Patricia Lynn.' (Martin)

Above and below: These two views of B-57Es show the high visibility caused by the brilliant orange paint scheme. As a safety measure, fighters would not fire upon the target banner until cleared by the tow target pilot when he could see that the fighter was locked on to the banner and not the tow aircraft.

mission and two hours was the maximum target towing range-time for the B-57E. Consequently, to save weight, two of the four reels were usually removed.

This tow target model of the Canberra was initially distributed to six units for target towing throughout the world. Due to the nature of air-to-air weapon training requiring a large amount of air space, practice areas were few. The B-57Es were assigned at bases close to these large restricted areas, and fighter interceptor squadrons deployed to these air-to-air ranges for this type of 'hot-fire' training. The first unit to receive B-57Es was the 17th TTS at Yuma AFB, Arizona, for use on their large air-to-air firing range. The squadron later moved to MacDill AFB, Florida, where this type of training continued for East Coast fighter interceptor squadrons. Service for West Coast fighter interceptor squadrons was then supplied by the 3rd TTS formed at George AFB, California. The 1st TTS operated from Biggs AFB, El Paso, Texas, while the 4756th TTS was based at Tyndall AFB, Florida to service the Fighter Weapons Center located there.

For overseas training, six of these new airplanes went to the 6th TTS at Johnson AB, Japan in mid-1957. Two airplanes and crews deployed for two weeks at a time to fighter squadron locations at Clark AB, PI; Andersen AFB, Guam; Naha AB, Okinawa; and Itazuki AB, Misawa AB and Yokota AB, the last three in Japan. The 6th TTS was inactivated by late 1957 and the B-57E resources were assigned as a flight of the 8th Bomb Squadron, 3rd Bomb Wing at Johnson AB. Nearly a squadron strength of B-57Es were stationed at Wheelus AB, Tripoly (Libya), where USAF European based fighter squadrons deployed for 'live firing' over their vast desert range.

To provide an ever increasing challenge for fighter intercepts, some B-57Es were modified to tow the styrofoam, bomb shaped, radar reflectant target. These could be towed higher and faster than

When firing passes on the banner were concluded and scored by the accompanying chase-plane, the cable was reeled in, bringing the banner to within 50ft of the aircraft as calculated by the cable counter. At this point a hydraulic cable cutter is activated and the banner would fall free, usually into the ocean below the firing range.

the high-drag 9ft x 45ft banner, but hits could be scored on the latter. By 1960, the 2.75in Mighty Mouse firing F-86D Sabre, F-89 Scorpion and F-94C Starfire were giving way to F-102 Delta Dagger with its six heat-seeking Falcon missiles. This made target towing for interceptor fighters obsolete, and the mission was ended for these 'Es.' A variety of new work was found for these airplanes, mostly by continuing to service the fighter units as ECM and faker target aircraft.

Although the 'E' was built with the capability of having combat equipment added if necessary, few of us ever thought this would become a reality. With the heavy attrition rate in Vietnam caused mostly by the two major ground explosions, 12 'Es' and eight 'Bs' were gathered throughout the Air Force and returned to the Martin plant in late 1965 to be combat configured (see Appendix 7). Consequently, when the war time need arose, the 'Es' served not only as tactical reconnaissance aircraft, but filled the role of light bombers as well.

25

FRIENDLY ENEMY: DSES

Since the B-57 became operational, it had been used as the most practical and realistic 'faker' or simulated target aircraft for interceptors throughout its career. The explanation was simple; reasonable speed, realistic high and low altitude capability, and sufficient range to begin an inbound track beyond the reach of intercept controlling radar.

In time however, the challenge to the defending fighter pilots and ground control intercept radar (GCI) had to be increased over that of merely detecting and attacking a normal radar reflective target. A would-be enemy would make the problem more difficult. This required special electronic equipment for the target aircraft. A number of Canberras were modified for this special mission beginning around 1959. Initially, RB-57As from inactive reconnaissance units were modified by Martin to have their bomb bay doors refitted to carry the latest electronic counter-measures and warfare systems to confuse the defenders. After first using air driven generators in the bomb bay to accommodate the added electrical demand, two of the four engine driven direct drive AC generators were replaced by Sunstrand constant speed units which account for the larger air scoop under the engine air inlet. The remaining two generators continued to provide AC power for the basic air frame in normal operation. In the event of generator failure, power from the alternators was available for all functions. Wing racks designed for bombs, now carried chaff dispensers, and duties of the navigator were replaced by an electronics warfare officer (EWO). As other models of the Canberra became available, the RB-57As gave way to 'Bs' and 'Es' for the advantages they had over the earlier 'As.'

Considerable realism could be generated into these simulated aggressor attack mission of the type flown by these Canberras. Surprise to the defenders was the keynote for the B-57 crews. Their unannounced take-off from a nearby field for a typical 'faker' mission was often in the very early morning darkness or any other time when least expected. The protecting radar scopes may have reveal their departure, but once outside the range of their surveillance, the EB-57s (E for Special Electronic Installation) would turn on a tangential course and return as a new and unidentified inbound track. Often times, many EB-57s were used to form separate tracks and provide a coordinated electronic jamming attack to complicate the problem. When inside the range of the radar and in anticipation of interception, chaff was dispensed to confuse the defence force, and

When Canberras took on a new role as an Electronics Systems Countermeasure aircraft, this colorful photo of an EB-57A was used for USAF promotional posters to depict one of many Air Force missions. A ground radar intercept station is on the mountain peak in the background. (USAF)

The side entrance hatch to the cockpit for the EB-57A made easy access, but unable to open the canopy, heat was trapped in the cockpit, raising temperatures to near intolerable conditions. Ground cooling units were a must in high ground temperatures. (USAF)

At the time that RB-57As were being phased out of reconnaissance squadrons, a new mission was at hand. Most were modified to perform the Defense Systems Evaluation duty as targets for fighter interceptor training. These Canberras bristled with antenna and had a bomb bay door packed with electronic gear. (USAF)

With both engines at idle and making an unforgettable throbbing beat, this Canberra is about to move away from the chalks for a mission. The wide canopy for only one pilot gave plenty of unrestricted movement, but visibility was far better from the later tandem and narrower canopy. (USAF)

This EB-57E flies wing on a F-102 back to Elmendorf AFB, Alaska, after a friendly intercept training mission. On 13 June 1969, this Canberra was accidentally rammed by an F-102 during a normal training intercept. Crews of both planes parachuted safely, but separate helicopters recovered the crews in case one or both were no longer friendly. (N. Taylor)

This ramp scene at Hill AFB, Utah shows the line of EB-57As belonging to the 4677th DSES assigned there. This snow patched ramp was typical of the conditions of which DSES crews and aircraft were frequently called upon to operate from. When the B, C and E models became more available from bomb squadrons, these A's gave way to the newer models. (Minert)

electronic pulses jam radar signals turned on. It was up to the defending interceptors and ground radar stations to sort out the correct information.

Units operating these specially equipped EB-57s were Defense Systems Evaluation Squadrons (DSES), in theory replacements for tow target units, giving better training against ground and airborne defence systems, realistically simulating an ECM supported attack. Throughout the 1960s there were a number of these units based around the US and a few overseas locations. Eventually most were absorbed into two squadrons of the Aerospace Defence Command. The 4713th DSES in the north-east U.S. was stationed at Stewart AFB, NY, Otis AFB, Mass. and finally Westover AFB, Mass. Aircraft of this unit were often seen in Europe in support of USAF fighter activities. The other unit was the 4677th DSES originating at Hill AFB, Utah, and later moved to Malmstrom AFB, Montana. Its area of interest was primarily Fighter Interceptor Squadron (FIS) training for units guarding the Canadian approaches to the US. In 1974, the 4713th DSES at Westover was inactivated and the aircraft were divided between two Air National Guard units, and the 4677th DSES was redesignated the 17th DSES. Operating EB-57Es exclusively, this unit was inactivated in July 1979 and was the last to fly Canberras in the USAF. It shared the Defence Systems Evalu-

This colorful picture of this EB-57E of the 4677 DSES shows it flying over the Great Salt Lake. This unit was based at nearby Hill AFB and later moved to Malmstrom AFB, Montana. (E. Bruch)

The Air Defense Command which operated the DSES units made good use of WB-57Ds when they were released from other duties. These provided very high targets coupled with ECM jamming, making for a very difficult intercept. These two EB-57Ds of the 4677th DSES are on their home base ramp at Hill AFB, Utah. (USAF)

The 17th DSES was the last active Canberra unit in the USAF, and operated EB-57Es exclusively. Their all grey aircraft with red tail, nose, and wing tips, have red, white and blue fuselage bands, shown here with the bicentennial insignia of 1976. This later gave way to the unit insignia. On the chaff dispenser is painted '17 DSES' over blue, followed by red and white stripes. Bulge on top and bottom of fuselage waist is a TACAN antenna seen on most models. (S. Nicolaou)

Preserving the tradition of the early B-57s that emerged from the factory in over-all black scheme, crew chief Geo. Boucher of B-57C, 53-3856 of the 158th DSES influenced this repainting to its original scheme. Somewhat symbolic of antiques in service, this airplane was among the last B-57s to be retired from military service after 26 years. (D. Menard)

ation mission with the Kansas and Vermont Air National Guard units, those being the 190th DSEG at Topeka, and the 158th DSEG at Burlington, both flying EB-57Bs. In 1978, the 190th phased into another aircraft and mission, and the 158th DSEG became the last U.S. military unit to fly B-57s. While operating as the last to fly the EB-57s as well, the 158th DSES supported ADC fighter squadron training and all major NORAD exercises throughout the U.S. and Canada.

This ANG squadron, the 158th DSES, began giving up their Canberras in October through December 1981, the last two departing Burlington, Vermont, on December 14, 1981 for desert storage at Davis Monthan AFB, Arizona. These were both 'C' models; 53-3831 and 53-3856. The pride that this unit had in their Canberras was reflected in the way they had these two B-57Cs painted in their final years of service. They were "restored" and operated in their original all-over black paint scheme just as they had been when they left the factory in 1955, over 26 years before their retirement. It was this Air National Guard unit that participated in this final chapter in the military operational history of the B-57 Canberra.

Similar missions were being flown in other parts of the world. This EB-57E and others belonging to PACAF and based in the Far East operated in SEA camouflage since they had a tactical mission as well. Three of these airplanes were assigned to the 6091st RS at Yokota AB, Japan, later redesignated 556th RS and moved to Kadena AB, Okinawa. (Picciani)

26

Air National Guard

The Air National Guard was the user of the B-57 from early 1958 when regular Air Force units began turning in their Canberras for other equipment. This association with the Canberra lasted continually until 1981 when the last militarily used B-57 was flown to storage from an Air Guard unit, the 158th at Burlington, Vermont. The initial role for the B-57 within the Guard was that of passive reconnaissance over the United States following any major disaster, using RB-57As. As other B-57 Air Force units phased into other aircraft, their Canberras were transferred to ANG units. This grouping of photos shows a representative aircraft from each of the seven state Guard units that used B-57s.

Air National Guard units operated B-57s for many years in diversified missions ranging from reconnaissance, ECM, to night interdiction. This reconnaissance converted B-57B of the Kentucky Air Guard shown in 1964, later served in Vietnam and is now in the National Air and Space Museum collection. (C. Arrington via R. Conely)

An early receiver of the B-57B in the ANG was the 154th Tactical Reconnaissance Squadron, Little Rock, Arkansas. These Canberras received in late 1957 were modified within the unit to carry one camera, but they soon gave way to RB-57As that were more suited for the reconnaissance mission. (R. Dorr)

(1) Michigan ANG RB-57A, 491 of the 172nd TRS, 110 TRG at Battle Creek, October 1962. The state name abbreviated was carried in standard letters on the nose of each ANG aircraft.

(2) The Virginia ANG's 149th TRS at Richmond was to convert from B-26s to Canberras in 1958. They received one RB-57A, 475 in April and it departed in June since that unit converted to F-84s instead. (F. Hartman)

(3) The Arkansas ANG had the most colorful markings for their aircraft within the Guard Bureau, using a lavish chrome yellow nose and fin flash. Shown is RB-57A, 440, of the 154th TRS 189th TRG based at Little Rock in 1962.

(4) RB-57A, 428 of the 117th TRS, 190th TRG at Hutchinson, showing the change to 'U.S. AIR FORCE' on the nose with the state name KANSAS on the fin and the ANG national insignia which became standard marking around May 1962. (D. Anderton)

(5) The Kentucky ANG gave up this RB-57B, 589 of the 165th TRS, 123rd TRG at Louisville, in January 1965 for combat in SEA. It crashed on take-off 30 June 1965 at Da Nang. Crew escaped.

(6) The Nevada ANG based their Canberras at Reno with the 192nd TRS, 152nd TRG. This B-57C, 831 photographed in November 1962, was one of 20 refitted for combat in 1965. It was held in reserve in stateside units and became the last of two B-57s to be retired from military service.

(7) The Kansas ANG absorbed B-57Gs in May 1972 of the 13th TBS from Ubon AB, Thailand. Assigned to the 117th TBS, 190th TBG at Forbes AFB, Topeka, they were retired 2 years later due to excessive systems maintenance problems. This photo taken at Forbes 2 June 1972 shows FK tail code retained from 13th TBS. (F. Roos)

(8) In early 1974, two ANG groups assumed the Defense Systems Evaluation role, one being the Vermont ANG with the 134th DSES, 158th DSEG at Burlington. EB-57B, 505 shown here in 1977, painted gray and red, carries VERMONT on the rudder. The name of this Guard unit, 'The Green Mountain Boys' painted on the chaff dispenser pod perpetrates the reputation of the spirited irregulars of the Vermont Militia dating back to the American Revolution. (W. Lavery)

(9) When the Kansas ANG retired their B-57Gs, in 1974, they became the 117th DSES, 190th DSEG equipped with EB-57Bs, one example being 551 shown here in mid-1975. This Canberra served many mission requirements during its operational service. See lead photograph that also depicts this aircraft.

(10) Reminiscent of the mid-1950s when Canberras came from the Martin factory painted black, the 134th DSES repainted their B-57Cs in the summer of 1977 to look like new. It was an attention grabber! This Vermont squadron was the last unit to operate the B-57 Canberra in the military role. (D. Menard)

27

Special Projects

Some airplanes are naturals for adaptation to various types of special projects. The C-47, and C-130 in their many configurations are but two. The B-57 Canberra without a doubt falls into this category of being an air vehicle for endless functions. Bombardment, reconnaissance, target towing and weather sampling were designed into the basic airframe, but follow-ons such as VIP transport, engine flight-test bed, radar mapping, storm plotting, systems flight test vehicle, armament systems, to name a few.

One requirement once placed upon the Canberra in the mid-1960s was to carry a missile probe to 50,000ft then to pitch up to a 90 degree angle to launch the missile vertically! The Canberra was a good airplane, but that was a bit beyond its capabilities. Compro-

Up to 13 B-57s were assigned at Andrews AFB between 1959 and 1962 for special military administrative flying. Two B-57Es painted all white with day-glo stripes, were equipped with Lear autopilots and assigned to Project Flagstaff for the Office of the Chief of Staff of the Air Force. This airplane was later modified to a RB-57E with a TFR (Terrain Following Radar) system for testing, then flew with Project Patricia Lynn in SEA.

The second of the eight B-57As, 52-1419, was operated by the National Oceanic and Atmospheric Administration of the Commerce Department out of Miami International Airport. One of its uses was in tracking hurricanes in Project Stormfury.

When the NOAA B-57A, N1005, was retired, it was acquired by the George T. Baker Aviation School at Miami International Airport for flight mechanic training. Externally, only the name changed.

The Air Force Special Weapons Center used this B-57B for flight testing and dropping various shapes. In this picture 498 is preparing to drop test a Jaguar air-launched sounding rocket over White Sands Missile Range, New Mexico. (USAF)

This specially configured B-57C was configured for use in Project Clean Sweep III for LASL and LRL in the early 1960s. This called for launching a data gathering rocket from a very high altitude and high pitch angle. (L. Davis)

This B-57B, 52-1576, was first to have 20mm cannon installed as a prototype for all those that followed. It was then consigned to NACA for flight test projects. Shown here at Cleveland International Airport, it is configured for Project Bee which was conducted by the Lewis Laboratory. This was experimenting with hydrogen as a jet fuel for high altitude flight.

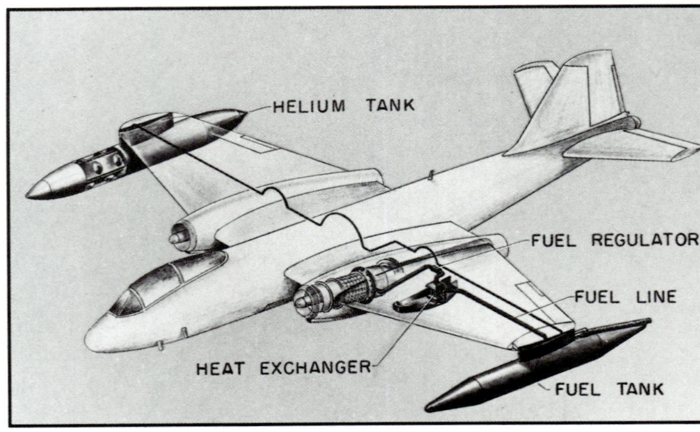

Schematic showing hydrogen system installation in B-57B 52-1576 for Project Bee.

mises were made, and the requirements of the test were able to be met by a B-57. The project was called *Clean Sweep III*.

For this special project, scientists from two weapons labs; Los Alamos Scientific Lab (LASL) and Lawrence Radiation Lab (LRL) wished to have samples of an air nuclear explosion from the earliest possible time and before the cloud had disappeared. The problems being faced were the extreme radiation levels after detonation and the extreme altitudes reached by the cloud.

During tests of Operation *Dominic* in 1962 on Christmas Island, special nosecones driven by Navy Shrike missiles failed to reach the desired altitudes of 100,000ft. The Shrike motor simply was not big enough.

An air launch would increase peak altitudes and was cheaper than redesigning the system. The lab asked if the B-57 could carry a sampler rocket and launch it at altitude in the maneuver described. This was not possible, but no one knew to what lesser degree this could be achieved.

The maneuvers selected had to be precisely repeatable while flying on instruments since test detonation was always timed for just before dawn. Weather at the test site was often unpredictable.

Several B-57s were modified with missile nose sections to air test missile guidance systems. This one, 52-1497, was modified by Timco to carry 17ft of the Boeing Bomarc. Separate hydraulic and electrical systems were added for missile guidance, and ammonia and nitrogen tanks were incorporated for cooling and pressurizing the missile nose. (J. Tarsitano via D. Menard)

Several missile guidance systems were tested including this TM-76A Mace missile nose section and components. In flight, the B-57 was guided by the ATRAN guidance system requiring the pilot only for landings and takeoffs.

Modified by Goodyear Aircraft Corporation, this B-57B is fitted with a TM-76A Mace missile nose. Much of the missile flight training for which these aircraft were configured was conducted in Germany, the area over which the Mace was programmed to be used.

Ideally, the maneuvers would also include an escape from the radiation in the cloud.

A Low Altitude Bombing System (LABS) [described on page 42] modified for the mission was installed in three B-57C aircraft. The launch angle was preset on the ground and the rocket that was carried under the wing, automatically fired at that angle. This allowed the pilot to concentrate exclusively on flying the maneuver.

D. E. Felts was one of the pilots for this project. Don describes two maneuvers that were developed. 'For launch angles of 55 degrees or less, initial altitude was 33,000ft. We would begin a steep dive at 45 to 50 degrees nose down with full power on the launch heading. The planned pull-up point was at 27,000ft, 5nm from the cloud, keeping wings level throughout the 'pull' which was maintained at 3G until the rocket fired. At that time, back pressure was relieved and the airplane was rolled 180 degrees to split-S from the nose high altitude. This also put the aircraft on a heading away from the cloud. The airspeed would drop off to 60 to 80kts, so great care had to be taken to avoid inducing a stall.'

The second type maneuver was even more spectacular! For rocket launch angles that were of a higher elevation than 55 de-

High altitude capability of the B-57 made them an excellent subject of optic and photographic evaluation. B-57B 52-1540 has a photo resolution pattern painted on its lower side for missile tracking optics used at the Kennedy Space Flight Center, Florida, in the early 1960s. (Dorrance)

To help with testing a laser camera in 1969, this NRB-57A of the RADC, 52-1435, was painted black underneath with photo resolution patterns painted in white. While flying at 40,000ft at night, the image of the aircraft more than filled the camera frame on the ground. The author was project pilot.

Another view of 52-1435 before being painted with a black undersurface. It met with misfortune in 1968 when its nose gear would not extend. Foamed runway at Griffiss AFB, NY prevented serious damage. Canopy was blown by pilot Larry Cunningham as standard measure for emergency landings of this type.

grees, Don Felts described this type of maneuver. 'The particulars for the maneuver were basically the same except the 'pull' was planned at 25,000ft, 3nm from the cloud. Back pressure of 3G with wings level was maintained as before, except this time the maueuver was carried "across the top" becoming a ballistic curve with zero airspeed at the top. After the nose came down below the horizon and flying speed was again attained, the power was reduced and a 180 degree roll (or what ever it took) put the aircraft right side up on a heading away from the cloud.' Don tried to recall what the airspeed may have 'maxed' out to be, but at that point of escape it couldn't be high enough!

Although only one data gathering rocket was fired into this cloud, it was radar-tracked to 108,000ft and seemingly everyone was happy with these results of *Clean Sweep III*.

Another extreme use for a B-57 was in testing liquid hydrogen as an aircraft propulsion fuel in order to increase operating altitude. If successful, this would allow aircraft to fly higher and out of the range of ground-to-air missiles, the major Soviet threat at that time. A higher flying airplane than the U-2 was the ultimate objective. The customary JP-4 fuel for jet aircraft had an effective operating altitude limitation around 60,000ft to 65,600ft. In simulated conditions at Mach 0.8 when using a wind tunnel and liquid hydrogen as a fuel, this altitude could be extended to 90,000ft, the limit of the facility. Such prospects of extending aircraft altitudes with liquid hydrogen became a joint effort from 1955 to 1959 between the power plant laboratory of Wright Field and NACA. It was known as Project *Bee*.

To prove the results of the wind tunnel tests in actual practice, the NACA Lewis Laboratory at Cleveland decided to use the B-57B already on bailment to them from the Air Force. It was the J65 engine on which ground tests were being run, and this airplane proved to be a natural. Presumably this was 52-1576. The basic plan was to equip the airplane with a hydrogen fuel system, independent of its regular fuel system, and modify one engine to operate on hydrogen as well as its regular fuel. The airplane was to take off and climb on its regular fuel. After reaching level flight at about 50,000ft, the fuel on one engine was to be switched from JP-4 to hydrogen. When the hydrogen experiment was completed, the fuel flow would be switched back to JP-4 and the airplane would land under normal operating conditions.

A number of B-57s were reassigned in 1962 for modification for nuclear air samplers. This was at a time when China was testing their weapons and debris from these nuclear tests were gathered over many parts of the world. Approximately six were fitted with two F-84F air pressurized drop tanks giving maximum fuel load for B-57s of 25,000lb (3,846 U.S. gal). Due to the added drag, they offered a mere 20 minutes air time, but made the westward trip to Hawaii less critical with headwinds. For ferry flights, standard tip tanks replaced sampler wing tip pods shown here. (G. Williams)

Photo resolution patterns were painted on the top of the wing of JB-57B, 52-1499, for use by cameras from satellites and balloon platforms in 1965. The B-57 could fly above the earth's haze layer for better unrestricted images under the cameras. (J. Andrews)

This strange nose modification on B-57B 52-1581 supported a project for night sensors and low light TV. A more refined system developed from this project was eventually used on B-57G Canberras for spotting targets in the dark. The airplane is seen here in storage at Davis Monthan AFB.

The conversion very basically consisted of two specially built wing tip tanks, one for the liquid hydrogen and the other for helium with which to pressurize the hydrogen tank to feed the engine. One of the J65 engines was modified so that a selection could be made to operate it on either JP-4 or the liquid hydrogen. The rest of the system included a heat exchanger under the left wing for vaporizing the liquid hydrogen, a flow regulator, and a manifold for feeding gaseous hydrogen to the engine.

Both wing tip tanks were very complex. The hydrogen fuel tank on the left wing of the airplane was 20ft 4in long with a volume of 1.7 cubic meters. The stainless steel tank was designed for a pressure of 3.4 atmospheres and insulated by a 2in coat of plastic foam, covered by aluminum foil and encased in a fiberglass covering. On the opposite wing was the helium supply tip tank containing 24 fiberglass spheres charged to 200 atmospheres.

Pilots for the experiment were William V. Gough Jr., a former Navy pilot then working for NACA. The second pilot in the rear seat was Joseph S. Algranti who operated the special controls of the hydrogen fuel system. The first flight was made on 23 December 1956. Upon reaching 50,000ft over Lake Erie (for the safety of those who would otherwise be below), the test was begun. While making transitions from JP-4 to hydrogen, the engine responded by overspeeding and vibrated heavily. The startled pilots quickly shut the engine down and purged the system. A safe single engine landing was made back at the Cleveland airport through a low overcase and light rain.

The second flight was also only partially successful. Transition from JP-4 to hydrogen was made successfully, but insufficient hydrogen flow prevented satisfactory high-speed engine operation.

On 13 February 1957, the third, and first of three successful flights was made and the fuel system worked well. The engine ran for about 20 minutes on hydrogen and the pilots found that the engine responded well to throttle changes – a very critical element. An interesting phenomenon was that the engine burning hydrogen produced a dense and persistent condensation trail, while the engine operating on JP-4 left no trail. Project *Bee* was considered a success whereby the feasibility of using liquid hydrogen for higher altitude flight had been demonstrated. However, flight experiments ended about here because requirements for such extreme altitudes for aircraft failed to materialize.

The early years of the B-57 was also a time of considerable missile experimentation. Guidance systems for these missiles was the prime factor for their success. In order to flight test these guidance systems, different types of missile nose sections with their guidance systems were built into the nose of several B-57s. These produced grotesque appearing airplanes, but the Canberras performed these air vehicle functions very well.

The Rome Air Development Center normally had three B-57s assigned for air test vehicles. They were always under some form of modification that would carry hundreds of different types of equipment being tested. These systems ranged from laser equipment, sensitive bomb dropping methods, photo mapping systems, to name but a few of the obvious. The B-57 could provide not only the speed and altitude as well as duration needed for these tests, but was an excellent stable platform that could be calibrated very closely for result verification. For tests of the nature described, there was no better aircraft in existence for the many years covered by the B-57.

The FAA acquired two RB-57As for their use in evaluating the high-altitude airways structure in the late 1950s. This was the beginning period of jet transports for domestic service and a new level of air navigation. (D. Menard)

28

FLYING THE B-57

Imagine this to be your familiarization ride in the Canberra. Our seat pack parachutes have already been positioned in the aircraft, so climb in the back seat while I make a walk around inspection and check the usual items from the checklist. Once this is done, our clearance filed, and we are strapped in, the hard work is over and what flying is all about is soon to begin.

Battery switch on. Now we can talk to one another on the interphone hot-mike. I've already gone through the 'before engine start' check list which assures that all the switches and handles are in the right place. The right-hand throttle is now moved out of the detent into the idle position and the right engine is ready for start. I signal to the crew chief for wind-up. At the front of the right console panel are two switches marked 'start' and I press the right-hand one up. Instantaneously, the one shot starter cartridge ignites to our right with a deafening sound like escaping high pressure steam. This starter is a small turbine that connects to the engine turbine and causes it to turn. This burning cycle lasts 10 seconds during which time the engine instruments come to life. Automati-

With seat packs and inflatable dinghies already on, the author gives Ron Cutting a helpful boost up the ladder of the B-57. Looking on is crew chief R. F. Frost. Place is Johnson AB, Japan, 1958.

This is the front office of the B-57. Very few switches, instruments or controls are considered hard to reach or see. The Canberra is a big airplane to require only one crew member, but it is an easy bird to fly. Arm rests are missed from the old type seats which were deleted on this Douglas ESAPAC seat modification.

Right pilot console. Above left to right: Four generator switches and voltmeters, TACAN and oxygen regulators, transponder (IFF), radio mixer and light switches. Below; oxygen quantity gauge, cabin pressurization and temperature control, UHF radio and monitor switches, and partially hidden by early style ejection seat armrest is the automatic direction finder (ADF). (D. Beggerly)

cally, ignition and fuel take over to bring the engine up to 42%, idle speed. What could be simpler? We always watch for a possible overheat, but I have never seen this happen with the J65 engine – one of the most reliable engines I have ever operated. A check of the hydraulic pressure – it is up and I signal for the crew chief to pull the landing gear down-lock pins which are then stowed in the right side fuselage access hatch. Be sure your hands are clear of the canopy rail, for the canopy is coming closed. When down and the red warning light goes out indicating it is locked, it becomes refreshingly quiet in the cockpit. The left engine is started like the first. Smoke from the starter swirls outside the canopy giving good reason why this engine is started after the canopy is closed. Both starter exhausts eject smoke out the right side of the engine nacelle.

Setting the throttles at 50% reduces the ground noise of the annoying, pulsating rumble-beat which seems unique to J65 engines. The thumbs-out signal to the ground crewmen is given and chocks are pulled. We remove our seat safety pin to arm it for emergency ejection. Ground Control clears us to taxi and we are on our way. Adding just a little power, our Canberra rolls forward and we tap the brakes as a functional check. Since steering is accomplished by differential braking, adding power to the engine on the outside of the turn eases us out of the confined parking area with no difficulty.

When in position on the runway, the throttles are pushed forward to their limits and the engines wind up quickly with a deafening roar to full RPM. There is a feel that the Canberra is ready to go as vented oil from the engines begins to swirl like smoke around the outside of the canopy. A quick check of all the instruments is made as brakes are released smoothly. The loud noise of the engines slips behind us as we gain speed rapidly, causing an almost quiet, eerie state. The rudder comes into play quickly, replacing the brakes for directional control. As we pass 100kts, our forward pressure on the husky control wheel is slowly changed to back pressure and the nose wheel comes off the runway. We hold this level atti-

Before dual control B-57Cs were available to the units, pilot checkouts consisted of one back seat familiarization ride, then move to the front seat and it was solo from then on. 'Cs' were not long in coming however, and the USAF acquired 38. Their combat potential remained equal to that of the 'B.' (Martin)

It was hardly noticeable, but the rear seat in all models, including this dual control B-57C, had the rear seat slightly off set to the left. Forward visibility for landing was fairly good from the rear seat.

Martin presented promotional lapel pins or tie tacks when they were closely involved with the B-57. The one on the left was an all purpose pin for selected receivers. The center pin was presented to pilots as they checked out in the airplane, and the one on the right was for pilots when reaching the 1,000 hour mark. No further recognition was made.

Martin presented these chrome display models of the B-57B to people involved with the airplane, particularly initial aircrews and senior staff. This type of gift from a vendor to government consumers became illegal beginning in the mid-1950s and many of these models had to be withdrawn.

tude as speed builds. As lift-off speed of 133kts is approached, the nose is raised higher and we are in an attitude to leave the runway in less than 4,000ft (based on 45,000lb take-off weight). The moment of lift-off is hard to detect, mainly because the large tires dampen the beat of the runway expansion joints. Unwary pilots have been known to retract the wheels too soon, thinking they were airborne only to have the airplane settle down on its belly. In the initial climb there is an illusion that the nose is still slightly down due to the negative slope of the side canopy rail.

Our initial climb is established anywhere between 250 and 360kts, depending on our mission. As we go through 10,000ft we note that cabin altitude has stopped climbing and it should remain there until we reach 20,500ft. From then on it will maintain a constant 3.5psi differential whereby when reaching our optimum level off altitude of 37,500ft, our cabin altitude will be 21,000ft. This is a comfortable pressure, but we still have need to keep our oxygen masks in place.

At level off altitude our cruise is 0.73 Mach giving us a true airspeed of 420kts. About 93% RPM will initially hold this speed. The airplane trims up well but it does take some watching to hold it on altitude for it tends to make a slow hunt longitudinally. At any altitude, the B-57 could outmaneuver most – if not all fighters of its prime operational time period. This was a result of its light wing-loading which permitted tighter turns than fighters with higher wing-loadings. Control forces are heavy at low altitudes since it does not have boosted controls. But a little grunting and groaning on the part of the pilot will make the airplane do what he wants it to do. It has a very positive feel while doing aerobatics such as loops, tight turns, rolls, Immelmann turns, etc. One can easily forget that the Canberra is supposed to be a bomber.

Ten minutes or more before starting descent from a prolonged stay at the cold, high altitude, the canopy defog switch must be turned on. It is easy to be caught short by an early descent before having a chance to preheat the canopy. That is why I carry an automobile windshield frost scraper in my leg pocket so I can clear the windshield on the inside – and see to land.

The landing field determines the type of pattern to be flown. Since we are landing at a tactical base we take the option to make a 360deg overhead approach. Our entry is 1,500ft above the ground and lined up with the runway at least two miles out. Speed is stabilized at 250kts (later 300kts) indicated. At the approach end of the runway, things begin happening fast, so hang on to your hat!

Simultaneously, the throttles are brought back to 55%, speed brakes out, and a snappy left break is established. This is a 60deg bank which is held until a 180deg turn is completed and puts us on downwind leg. Airspeed has dropped to 200kts which is gear lowering speed. The gear handle is placed to the down position, and a slight jolt is felt as the heavy wheels drop out of their wells. After a short rumble they snap into place with a positive feel and we check the indicator lights and the sound of an unsafe gear horn, and we confirm that the gear is locked down. This brings the speed down quickly to 170kts so the speed brakes are brought in by the thumb switch on the right throttle, and the flap switch behind the throttles start the flaps down. Back pressure and a few clicks of the thumb switch on the control wheel for stabilizer trim holds the nose up for a few moments more until reaching 150kts, and a glide is established by lowering the nose.

'Langley Tower, Randy 34, turning base, gear check complete.'

'Randy 34, cleared to land.'

This turn will be a continual one on around to line up with the runway. At the base leg point, airspeed is reduced to 140 and rolling out on short final we are looking for 130kts. We haven't needed power since retarding the throttles to 55% at the break.

On final we are just a little high. Left rudder and right aileron establishes a comfortable slip and altitude is adjusted right where we want it. Over the fence at 110kts and round-out is begun for setting the Canberra right on the numbers at about 98kts. (Based on 32,000lb landing weight.) Control pressures are moderate provided back pressure is trimmed out. The big tires and long strut travel makes the B-57 sit down with a soft touch. When on the runway, the nose wheel is lowered slowly and it settles closer to the ground than at take-off. All the remaining fuel is in the forward fuselage tank now, which compresses the nose wheel strut more than when

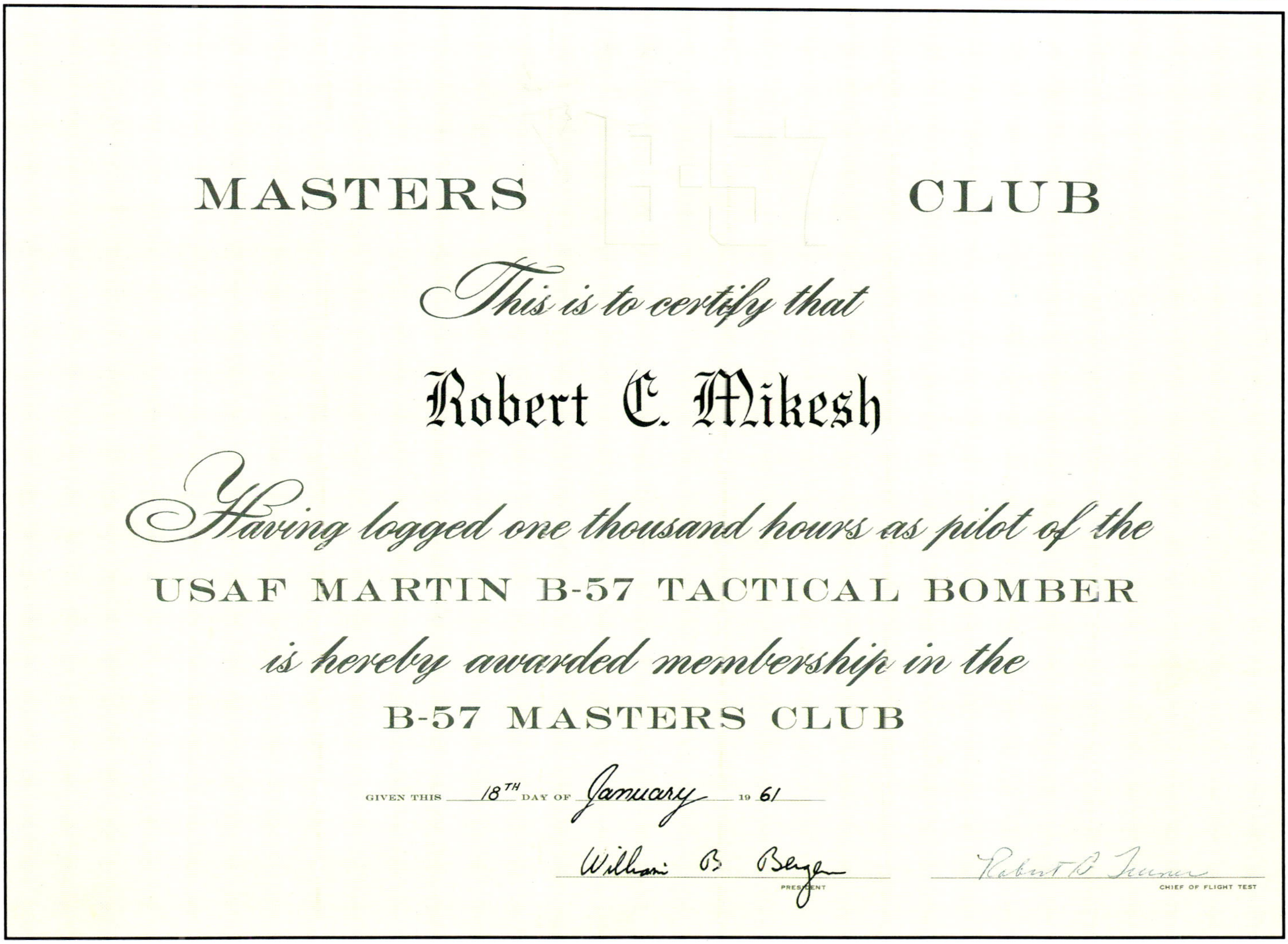
MASTERS B-57 CLUB

This is to certify that

Robert C. Mikesh

Having logged one thousand hours as pilot of the

USAF MARTIN B-57 TACTICAL BOMBER

is hereby awarded membership in the

B-57 MASTERS CLUB

GIVEN THIS 18TH DAY OF January 19 61

William B. Bergen
PRESIDENT

CHIEF OF FLIGHT TEST

In addition to the 1,000 hour lapel pin, pilots achieving this milestone were also presented these certificates. The airplane remained in service for so many years that many pilots far exceeded this mark. Martin ended this 1,000 hour recognition around 1959.

all tanks are fully serviced. Roll out can be near 2,000ft if we planned it that way – but we will spare the brakes and roll to an easy turn-off. Flaps coming up. (There are two flap positions; full 'down' which is 60deg, and 'up'.)

This power-off pattern was a little tighter than considered normal, but it was a procedure I liked to follow for getting the airplane down in the shortest length of time. Throttle adjustments were not necessary and as in the case with flying light aircraft, the runway could be reached in a glide at any point in the pattern.

If this flight sequence sounds over simplified, in truth it is not. The B-57 was an easy and operationally comfortable airplane to fly – and it was fun. It was docile and forgiving – but it did have limitations that if exceeded could bring trouble. A word of caution that I recall was said by 'Pat' Tibbs, Martin's Chief Test Pilot for the B-57 when telling us about this airplane, kept me alert in the Canberra at all times: 'Anything that will disintegrate on impact with the ground – can kill you.'

29

THE AIRPLANE

There is a personal side about the B-57 that cannot be found in flight manuals or performance curves. What the aircraft can really do is only known by the men that flew the Canberra and understood it intimately for the better part of their Air Force careers. So that I would not impart only one pilot's opinion, I queried many of my colleagues for their viewpoints on selected aspects about the airplane. Their responses often varied, but collectively they provide a good cross section for a pilot's perspective of the Canberra.

When asked what they liked most about the B-57, the list became long. Adjectives heading the list of qualities were: versatility, reliability and simplicity in both operation and maintenance. It was not one of the fastest airplanes for its time, but this was countered by its flight endurance, range, and stability in both level flight and as a gun and bombing platform. The Canberra's maneuverability at all altitudes and its twin-engine reliability were strong features mentioned by most pilots. Its cockpit layout was liked by all. Doug Beggerly summed up his opinion of the B-57 as being a 'lovable old thing – comfortable as an old shoe.'

The dislikes about the B-57 made a much shorter list than its good features, and some of those questioned even left this space blank. Poor single engine control at low speed of 155kts minimum (135kts for B-57E and others modified later with rudder power boost) was the most common criticism. A deadly mistake was to allow the good engine to surge to full power at or below this speed, which then exceeded rudder control capability. This took a lot of lives. Some wished for boost on all the controls but recognized that the airframe could have been easily over stressed by overzealous pilots. Canopy fogging often got ahead of the de-fog system and should have been improved. And when it came to replacing spent starter cartridges and cleaning their electrical contact points after each flight nothing could be grimier.

During peacetime tactical flying was one training mission after another. Following a practice weapons delivery sortie at Mito Range, Ellis Bruch (left) with navigator Mike Michaud, discuss their flight as they walk from their B-57 at Johnson AB, Japan in 1957. Back then, AF-blue flying suits were popular. (E. Bruch)

A goodbye between friends! After 3,633 flying hours, mostly for flight evaluation purposes, this NRB-57A, 435 with modified nose was retired in December 1969 to the aircraft reclamation depot and eventually scrapped. It was a nostalgic flight for both plane and pilot, the author, who flew this Canberra on its final trip to the bone-yard.

This was the Military Aircraft Storage and Reclamation Depot in the desert at Davis Monthan AFB, Arizona, which is now called AMARC (Aircraft Maintenance and Regeneration Center). Unless placed here for storage and possible future use, they are stripped of parts to keep others flying, and are eventually scrapped. Except for those assigned to museums, this was a sad ending for these Canberras after they had served their time so faithfully.

Performance charts say one thing but practical application is another. Some of the greater distant fights (and most memorable) were the frequent – into the wind and no room for error – flights from California to Hawaii (2,160nm) en route to the Far East. Equipped with ferry tanks, they always went over five hours. Bob Lince for instance, logged 6hr 5min in March 1966, to reach Hilo, the closest field short of Hickam AFB, the usual landing point. He calculated 6hr 26min to dry tanks. Other critical overwater flights were from Johnson AB, Japan, to Clark AB in the Philippines (1,650nm) as well as Guam (1,375nm) without ferry tanks, took about 4hr 20min and 4hr respectively. My logs show non-stop flights from Los Angeles to Washington, D.C. (2,050nm) ranging from 4hr 25min to 4hr 40min without ferry tanks, but we had to be assured of clear weather on arrival, and no delays. This left about 900lb of fuel showing on the gauge. Winds at altitude were always an important factor as well as having alternate airfields if fuel consumption was not as planned.

Time in the air depends on power setting and altitude for endurance as well as distance. For 'fuel on board' we normally filed having 4hr 15min which comfortably left 1,500lb remaining as minimum on cross-country flights. One of the longest flights without a ferry tank was 5hr 10min claimed by Bob Hunter from Reno, Nevada to Otis AFB, Massachusetts (2,250nm). This was in an RB-57A, and the lightest model of the B-57 series. A ferry tank would add another hour to flying time.

What the *book* claims about altitude is one thing and actual practice is another. Topping 50,000ft with standard wing B-57s became rare as the airplanes grew older and gained weight. When new and often without tip tanks, many topped it for one reason or another. A few worth mentioning are 53,500ft by Jerry Russell, 54,000ft by Jim Goodnight, 55,000ft by Bob Hunter, and 56,900ft by Doug Beggerly. This latter was during the Eniwetok atomic bomb tests in 1956 when the airplanes were lighter, but I do wonder what the tailpipe temperatures were reading, for this was often the limiting factor for altitude.

A handful of Air Force pilots have logged close to or over 5,000hr in the Canberra, and that is a lot of time in any military airplane. Paul Pitt logged a staggering 5,209hr in the B-57 before retiring. Another high-timer is Charlie Leonard who finished with 5,060hr, and may be the highest in the 'F' model with 1,808hr. He and Ted Jensen are the only two pilots to have flown all models of production B-57s.

The accomplishments of the Canberra and the men who flew her, can go on and on. No one book can contain all there is to be said for this fine airplane. The B-57 served quietly and modestly, seldom in the spotlight, and therefore may never be voted into the aviation hall of fame. But if the record of its accomplishments were to be examined closely, it must be considered one of the all-time great airplanes. Those of us who knew her, salute the few ships that are now only museum aircraft, for there will never be a true replacement for the versatile B-57 Canberra.

When the 17th DSES turned their EB-57Es in for storage at Davis Monthan AFB, much of this last squadron of B-57s from the USAF was set aside for needs that may arise. There was some consideration that they might be sent to the Chilean Air Force but that has now faded. Their fate will indoubtadly lead to scrapping.

Typical view showing one of 100 B-57s that ended their existence at AMARC following fully successful careers. While sad in appearance, these airplanes survived others for over 20-years that either crashed or were combat losses. This B-57C not only flew combat missions with the 8th Bomb Squadron, but as a dual control airplane, also transitioned new pilots into the Canberra. Seen here in retirement before being scrapped, it gave up many of its parts to keep other B-57s flying.

TYPICAL PILOT'S STATION FORWARD VIEW (AFTER RADIO SET AN/ARN-21 MODIFICATION)

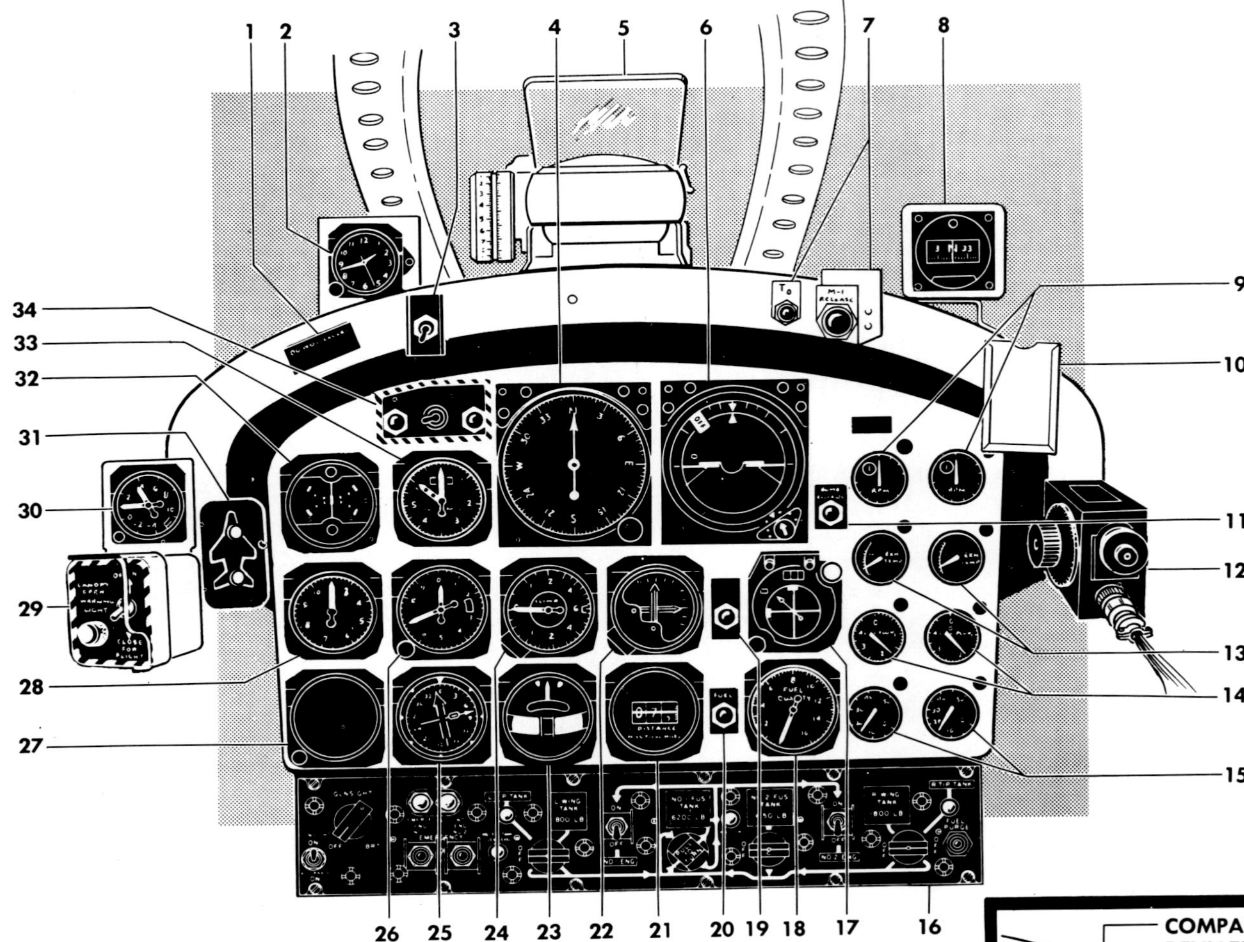

1. INDICATED AIRSPEED WARNING NAMEPLATE
2. CLOCK
3. HORIZONTAL STABILIZER EMERGENCY SWITCH
4. DIRECTIONAL INDICATOR
5. GUNSIGHT (REMOVED FROM RB–57B AND RB–57C AIRPLANES)
6. ATTITUDE INDICATOR
7. BOMB RELEASED INDICATOR LIGHTS
8. STANDBY COMPASS
9. ENGINE TACHOMETERS
10. COMPASS DEVIATION CARD
11. BOMB RELEASED INDICATOR LIGHT
12. POSITION SETTING UNIT
13. ENGINE EXHAUST TEMPERATURE INDICATORS
14. FUEL FLOW INDICATORS
15. ENGINE OIL PRESSURE INDICATORS
16. FUEL CONTROL PANEL
17. COURSE INDICATOR
18. FUEL QUANTITY INDICATOR
19. COURSE SETTING INDICATOR LIGHT (B–57C ONLY)
20. FUEL QUANTITY TEST SWITCH
21. RANGE INDICATOR
22. LABS INSTRUMENT
23. TURN AND SLIP INDICATOR
24. VERTICAL VELOCITY INDICATOR
25. RADIO MAGNETIC INDICATOR
26. ALTIMETER
27. FLIGHT COMMAND INDICATOR
28. MACH NUMBER INDICATOR
29. CANOPY ACTUATING SWITCH
30. ACCELEROMETER
31. AN/APS–54 WARNING INDICATOR
32. POSITION DEVIATION INDICATOR
33. AIRSPEED INDICATOR
34. FIRE EXTINGUISHER PANEL

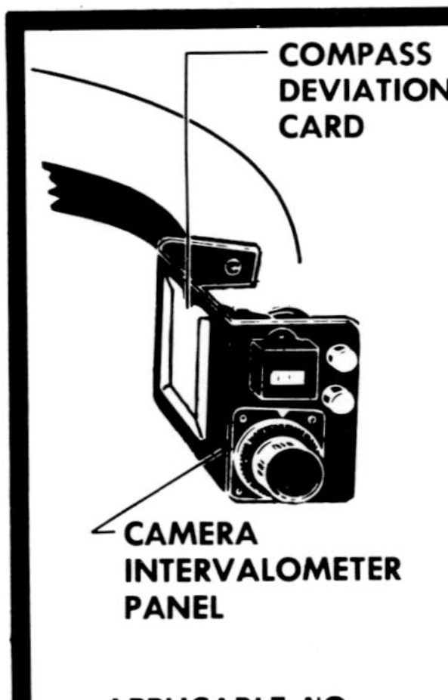

APPLICABLE TO RB–57B AND RB–57C AIRPLANES

1–50880A

Figure 1–29

TYPICAL PILOT'S LEFT CONSOLE

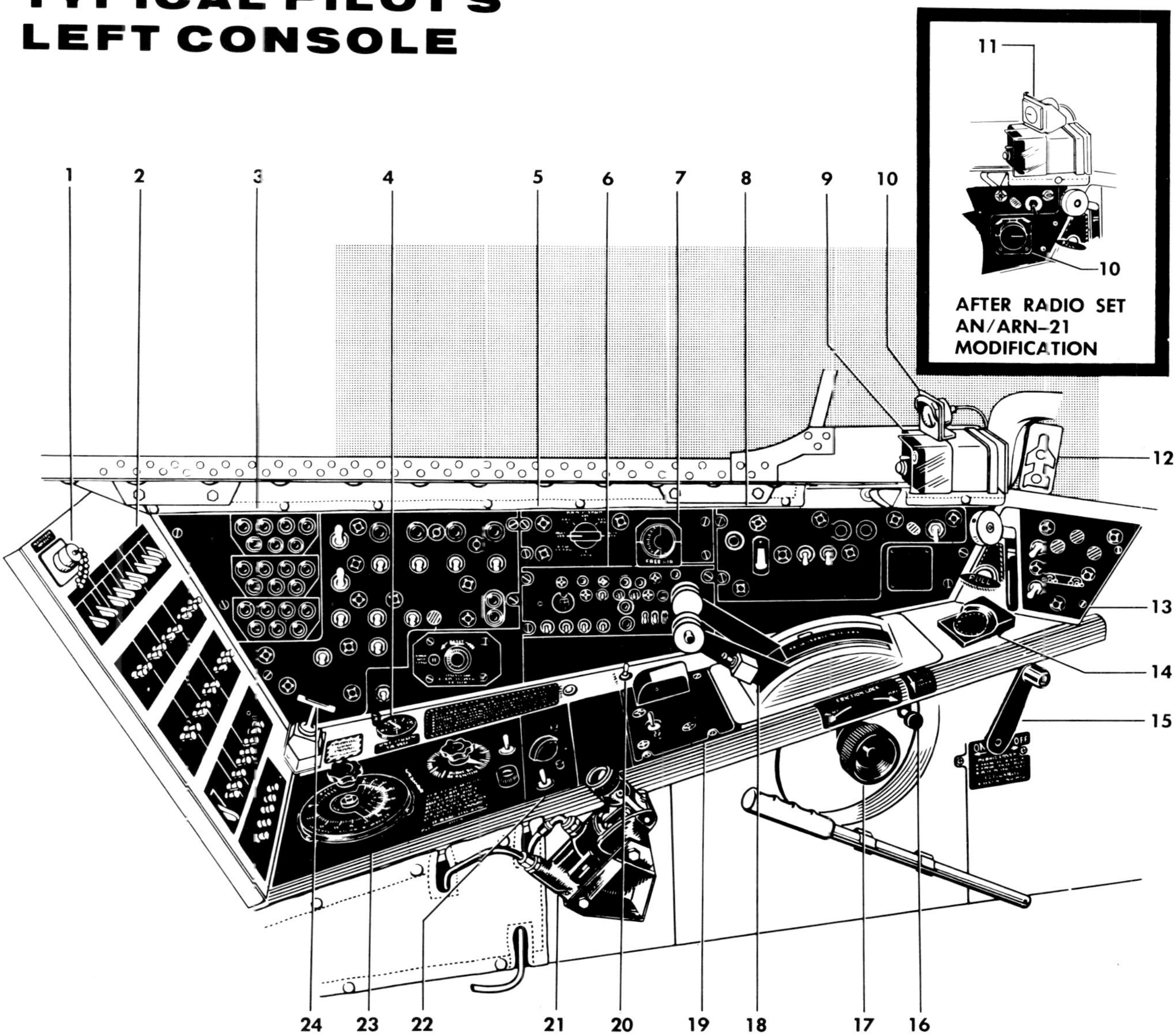

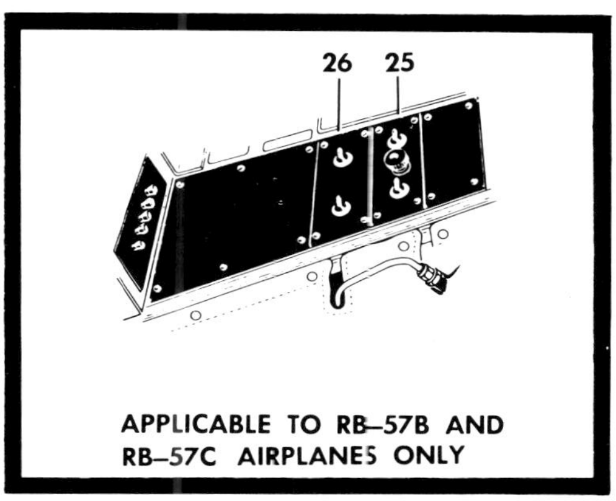

1. T-23 POWER RECEPTACLE
2. CIRCUIT BREAKER PANEL
3. ARMAMENT CONTROL PANEL
4. HYDRAULIC SYSTEM PRESSURE GAGE
5. SELECT ARMAMENT CONTROL PANEL
6. T-145 CONTROL PANEL
7. FREE AIR TEMPERATURE GAGE
8. PILOT'S LEFT MAIN CONTROL PANEL
9. CANOPY ACTUATING SWITCH
10. STABILIZER POSITION INDICATOR
11. ACCELEROMETER
12. RADAR WARNING INDICATOR AN/APS-54
13. LANDING GEAR CONTROL PANEL
14. BRAKE PRESSURE GAGE
15. PARKING BRAKE
16. WARNING HORN RELEASE
17. THROTTLE FRICTION KNOB
18. THROTTLES
19. FLAP AND TRIM CONTROL PANEL
20. J-2 COMPASS CONTROLS
21. HYDRAULIC HAND PUMP
22. LABS CONTROL PANEL
23. BOMB RELEASE INTERVAL CONTROL
24. EMERGENCY CANOPY RELEASE
25. CAMERA CONTROL PANEL
26. CAMERA VACUUM AND HEATER BLANKET CONTROL PANEL

1-12141C

Figure 1-30

TYPICAL PILOT'S RIGHT CONSOLE

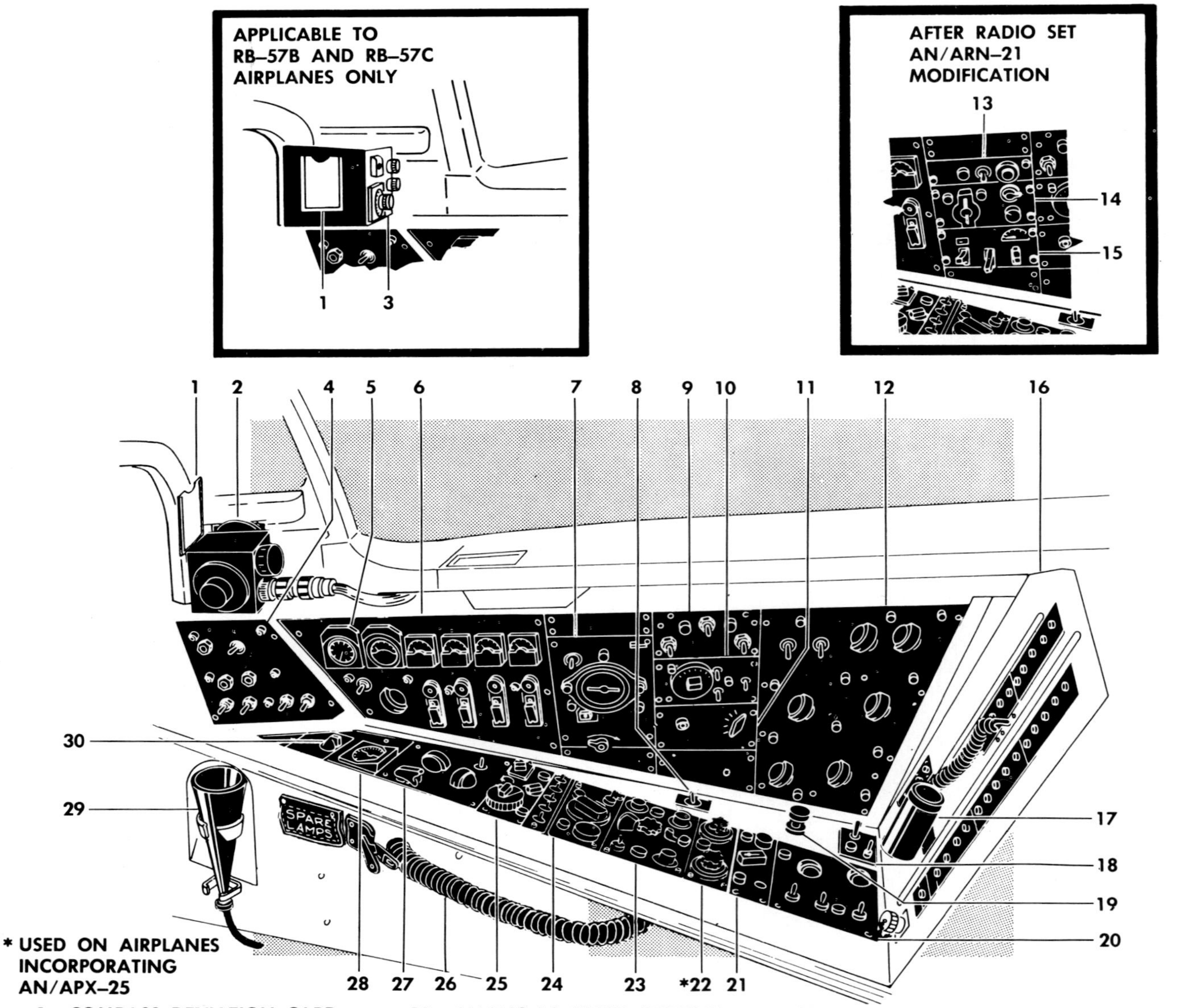

1. COMPASS DEVIATION CARD
2. POSITION SETTING UNIT
3. CAMERA INTERVALOMETER CONTROL PANEL
4. INVERTER CONTROL PANEL
5. CABIN PRESSURE GAGE
6. GENERATOR CONTROL PANEL
7. OXYGEN REGULATOR
8. MARKER BEACON–RADAR WARNING SWITCH
9. AUXILIARY LIGHTING CONTROL PANEL
10. IFF RADAR CONTROL PANEL
11. AN/AIC–10 FILTER CONTROL PANEL
12. PILOT'S LIGHTING CONTROL PANEL
13. TACAN SHIFT PANEL
14. TACAN CONTROL PANEL
15. OXYGEN REGULATOR
16. CIRCUIT BREAKER PANEL
17. C–4A LAMP
18. FERRY TANK CONTROL PANEL
19. DEFOG AIR KNOB
20. WINDSHIELD CONTROL PANEL
21. AN/APW–11 RADAR CONTROL PANEL
*22. IFF CODER CONTROL PANEL
23. AN/ARN–6 RADIO COMPASS CONTROL PANEL
24. AN/AIC–10 INTERPHONE CONTROL PANEL
25. UHF CONTROL PANEL
26. OXYGEN HOSE
27. AIR CONDITIONING CONTROL PANEL
28. OXYGEN QUANTITY GAGE
29. RELIEF TUBE
30. FUEL QUANTITY SELECTOR SWITCH

1–12143D

Figure 1–31

30

Combat Transport

As the first B-57s came off the production line in June 1954, Martin proposed to the Air Force a *Utility Courier* based on the basic Canberra airframe. A passenger compartment seating 6-11 extended forward from the main wing spar and added 24in to the length of the nose. Larger capacity center line wing tip tanks of 600gal each, fuselage tanks behind the wing spar of 915gal and additional tanks in the wings totalled 3,580gal (23,270lb), 140gal (910lb) more than a B-57B with ferry tank. Cabin pressurization of 8,000ft would be maintained to 40,000ft. Other interior configurations included ECM, bombardier-navigator trainer, photographic-reconnaissance trainer, cargo courier and facilities for litter patients. This proposal did not develop beyond the mock-up stage. Ahead of its time as we now know corporate jets, it was offered to the wrong market; military and not civilian. It would have been a large and expensive airplane in many aspects for the few passenger seats it could provide, but for those that ventured into this field, it would have been the first in the corporate class of jet transport.

Side-by-side pilot seating in a modified cockpit arrangement was part of the transport version proposal. Engines could have been upgraded from the standard J65s to J79s. The 600 US gallon centerline tip tanks shown here were seriously considered for the B-57B but never adopted.

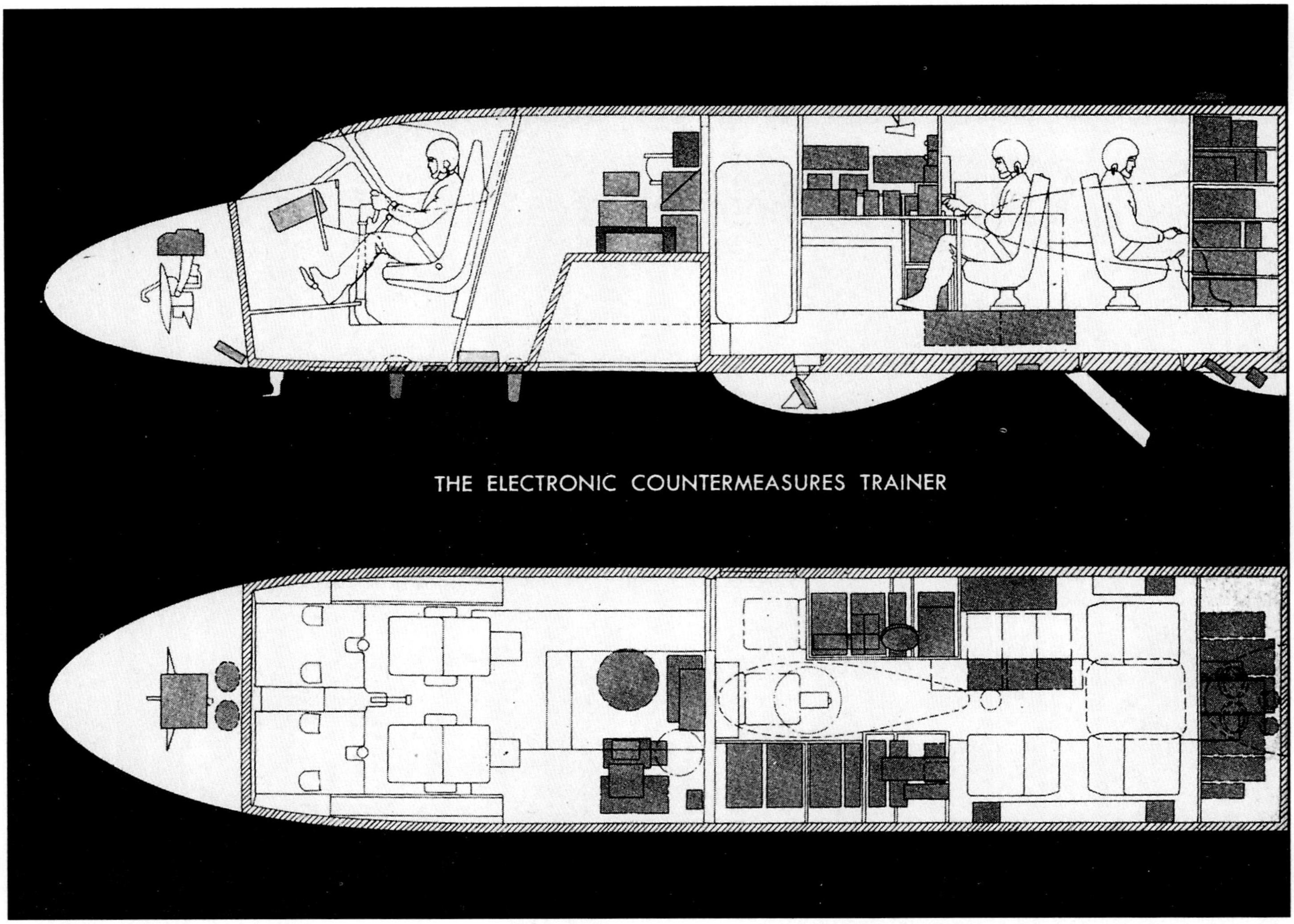

Above: Interior of this mock-up has four seats and two bunks. When configured with bench seats facing aisle, 11 passengers could be carried. A lavatory was opposite entrance doorway, and ample baggage space was in aft fuselage. Headroom in aisle measured 5ft 8in. (Martin)

Right: This mock-up of the B-57 transport shows cabin access just forward of the starboard wing. Movement of SAC alert crews from one site to another was big selling feature from Martin at the time. (Martin)

Appendices

Appendix 1
Performance Comparison

Model	B-26	AJ-1	B-45	Canberra	B-51	A2J-1
Gross Weight lbs.	36,300	51,023	95,960	48,421	64,154	56,000
Crew No.	3	3	3	2	2	3
Engine Model	R-2800-79	R-2800-447 J33-A-23	YJ65-W-1	YJ65-W-1	YJ65-W-1	XT40-A-6
Engine Rating	2000 BHP	2300 BHP 4600lb.	7220lb.	7220lb.	7220lb.	5035 BHP
Fuel Weight lb (Internal Tanks Only*)	6,978*	10,902*	36,100	17,973	24,700*	15,720*
T/O Ground Roll ft.	4,820	1,357	3,850	2,880	4,980	1,200
T/O Over 50' Obstacle ft.	5,710	2,400	5,300	3,600	6,350	1,560
Wing Loading lb/ft ^{2}	66.7	59.0	81.4	50.4	116.5	65.8
Load Factor	3.67	4.9	3.88	5.20	6.3	2.67
Combat Radius nm	800	800	845	990	800	995
Cruising Speed kts	199	199	400	412	458	363
Maximum Speed at Altitude kts	319	374	520	550	620	420
Max. Speed at Combat Altitude beginning at S.L. (Limited by Gusts*)	319	374	472*	434	595	400
Combat Altitude	S.L.	S.L.	S.L.	S.L.	S.L.	S.L.
Service Ceiling ft.	27,730	41,000	47,700	51,400	47,800	44,300
Radius of Turn & Speed at 3.5G	1000'-200kn	1065'-172kn	1150'-208kn	950'-175kn	2670'-297kn	See L/G
Approach Speed (120% Stall) kts	97	102	115	88	121	94
Stalling Speed at Landing Wt. kts.	81	85	88	73	105	79
Footprint Pressure psi	Main 69 Nose 70	Main 101 Nose 118	Main 167 Nose 160	Main 100 Nose 74	Front 65 Rear 78	Main 119* Nose 120
Landing Weight lb.	26,380	35,062	53,935	29,241	37,168	38,852
Landing Distance, Ground Roll ft. (Wheel brakes only)	2,000	1,742	2,860	1,300 max. brake	2,665	
Mission Time	8.22	8.32	4.17	4.98	3.65	5.74
Reserve at Long Range Speed	108	160	323	200	264	298
Target Action (minutes at long range at altitude)	44	50	30	38	47	24
Takeoff Rate of Climb fpm	1,300	2,530	7,750	5,700	6,700	4,250

APPENDIX 2
FACTS & FIGURES

Designer and Manufacturer: Basic design by English Electric Co, and The Glenn L. Martin Co, produced the American models.
Wingspan: 64ft.
Length: 65ft 6in.
Wing area: 960sq ft.
Weight: Empty weight about 27,000lb.
Fueled for takeoff about 45,000lb. Maximum weight 55,000lb.
Max speed: 444kts IAS with tip tanks 513kts IAS without tip tanks but not to exceed .78 Mach with tip tanks; .82 to .85 Mach is buffet area.
Normal cruise: About 420kts TAS, or .74 Mach.
Service ceiling: Above 53,000ft.
Propulsion: Two J65-W-5 or J65-BW-5 engines, each 7,220lb st at sea level.
Armament: Eight M3 .50cal machine guns with 300 rounds up to 52-1575. Four M39 20mm cannon with 290 rounds each from 52-1576 on.
Example of Stores Capability: *Internal* Four 1,000lb bombs, nine 500lb bombs, 21 260lb bombs, or two Mk 9 1,500lb Special Weapons.
On wing pylons Four 750lb stores or less.
Rockets Eight 5in HVAR, or 28 2.75in FFAR. Double if wing bomb pylons are utilized.
Ferry Range: More than 2,000nm with reserves using ferry tank, zero-wind.
Fuel (usable):

		U.S. Gal	*Pounds*
Fus Main	1	1,010	6,565
Fus Aux	1	662	4,303
Wg Outboard	2	580	3,770
Wg Drop	2	640	4,160
Total		2,892	18,798
Ferry tank		548	3,562

APPENDIX 3
PRODUCTION

Model	Serial Numbers	Quantity
Contract AF33(038)-22617		
Canberra	51-17352 (B2 WD940)	(1)
Canberra	51-17387 (B2 WD932)	(1)
B-57A	52-1418 to 52-1425	8
RB-57A	52-1426 to 52-1492	67
B-57B	52-1493 to 52-1594	102
Contract AF33(600)-22208		
B-57C	53-3825 to 53-3858	34
B-57B	53-3859 to 53-3935	77
B-57C	53-3936	1
B-57B	53-3937 to 53-3939	3
B-57C	53-3940	1
B-57B	53-3941 to 53-3943	3
B-57C	53-3944	1
B-57B	53-3945 to 53-3947	3
B-57C	53-3948	1
B-57B	53-3949 to 53-3962	14
Contract AF33(600)-25825		
RB-57D	53-3963 to 53-3982	20
Contract AF33(600)-29645		
B-57E	55-4234 to 55-4301	68
	Total:	403

APPENDIX 4
RB-57D VARIATIONS

Group	Serial	Model	IFR*	Crew
A	53-3977 to 53-3982	RB-57D	No	1
B	53-3970 to 53-3976	RB-57D	Yes	1
C	53-3964 to 53-3969	RB-57D-2	Yes	2
	53-3963	RB-57D-1	Yes	1

*In-Flight Refuelling

APPENDIX 5
GENERAL DYNAMICS RB-57F PRODUCTION RECORD

Serial	Constructed from	Model
63-13286	52-1559	B-57B
63-13287	53-3864	B-57B
63-13288	52-1539	B-57B
63-13289	52-1527	B-57B
63-13290	52-1562	B-57B
63-13291	52-1574	B-57B
63-13292	52-1594	B-57B
63-13293	52-1583	B-57B
63-13294	53-3935	B-57B
63-13295	53-3918	B-57B
63-13296	53-3897	B-57B
63-13297	53-3900	B-57B
63-13298	52-1536	B-57B
63-13299	52-1573	B-57B
63-13300	52-1427	RB-57A
63-13301	52-1432	RB-57A
63-13302	52-1433	RB-57A
63-13500	53-3972	RB-57D
63-13501	53-3975	RB-57D
63-13502	53-3970	RB-57D
63-13503	53-3974	RB-57D

APPENDIX 6
MARTIN B-57G CONVERSIONS

52-1578	53-3886
52-1580	53-3889
52-1582	53-3898
52-1588	53-3905
53-3860	53-3906
53-3865	53-3928
53-3877	53-3929
53-3878	53-3931

APPENDIX 7
MARTIN B-57S RECONFIGURED FOR COMBAT AT MARTIN FACTORY, LATE 1965

B-57B	52-2498	B-57E	55-4238*
	52-1499		55-4248
	52-1510		55-4251*
	52-1550		55-4259*
	52-1590		55-4265
			55-4268
B-57C	53-3827		55-4269
	53-3831*		55-4270
	53-3838		55-4274*
			55-4282
			55-4284
			55-4285*

*Indicates that these were not deployed to SEA

APPENDIX 8

MARTIN B-57S DELIVERED TO PAKISTAN, LATE 1959

B-57B	53-3885
	53-3891
	53-3938
	53-3939
	53-3941
	53-3942
	53-3943
	53-3945
	53-3946
	53-3947
	53-3949
	53-3950
	53-3951
	53-3952
	53-3954
	53-3955
	53-3956
	53-3957
	53-3958
	53-3959
	53-3960
	53-3961
B-57C	53-3834
	53-3846
	53-3948
Total:	25

APPENDIX 9

MARTIN B-57 UNIT TAIL CODES, 1967 AND ON

PQ: 8TBS, 405TFW, latter assigned to 35TFW

PV: 13TBS, 405TFW, Clark AB, Philippines

FK: 13TBS, 8TFW, Ubon, when equipped with B-57Gs

FS: 4424 CCTS, 15TFW, later 1TFW, MacDill AFB, trained B-57G crews

GT: 556RS, 347TFW, Yokota, later Kadena, EB-57Es

JM: 4461 TEWS, 363TRW, was 4416CCTS

JO: 22TRS, later 62TRS, 363TRW, Shaw AFB

APPENDIX 10
MARTIN B-57 COMBAT EFFECTIVENESS

To evaluate in true perspective, the effectiveness of any one airplane in combat, it must be compared to all participating airplanes. The following three tables reveal much by which to evaluate, but weapon delivery accuracy, and bomb damage assessment are difficult, if not impossible to record for true comparison. It was in this unrecorded area that the B-57 was recognized for its effectiveness.

Sortie Generation Rate in SEA
(Average daily sortie per aircraft; total sorties [minus training and maintenance] divided by daily average of possessed aircraft, divided by days of month)

	1968 *Monthly Average*	1969 *January to June*
A-1	1,441 / 59 = 0.79	1,573 / 61 = 0.86
A-26*	290 / 13 = 0.72	321 / 16 = 0.67
A-37	1,260 / 22 = 1.85	747 / 20 = 1.25
B-57	686 / 22 = **1.01**	337 / 10 = **1.12**
F-4	6,270 / 207 = 0.98	6,498 / 224 = 0.97
F-100	8,192 / 229 = 1.15	7,571 / 238 = 1.06
F-105	2,776 / 104 = 0.86	2,116 / 82 = 0.86
RF-4	1,849 / 69 = 0.86	1,912 / 74 = 0.86
RF-101	424 / 16 = 0.85	395 / 16 = 0.82

* As B-26K Invader when designation was converted back to the "A" attack series aircraft.

USAF Sorties in South Vietnam only

	1967		1968		1969	
	Total sorties	*Combat sorties*	*Total sorties*	*Combat sorties*	*Total sorties*	*Combat sorties*
A-1	3,712	3,000	4,042	3,166	2,621	2,055
A-26	0	0	0	0	0	0
A-37	0	0	15,033	14,447	8,761	8,305
B-57	**5,818**	**5,570**	**2,064**	**1,605**	**526**	**421**
F-4	20,284	16,392	25,381	23,134	21,965	19,185
F-100	83,782	80,398	92,612	88,250	57,561	52,699
F-105	0	0	2,072	1,811	0	0
RB-57	**1,153**	**1,017**	**1,073**	**1,021**	**1,110**	**1,058**
RF-4	8,540	7,996	9,173	8,719	7,221	7,039
RF-101	2,937	2,650	3,762	3,498	2,942	2,637

USAF Aircraft Loss Rate in SEA

	In South Vietnam 1 Jan 65-31 Oct 69			*In North Vietnam 7 Feb 65-31 Oct 69*			*In Laos 18 May 64-31 Oct 69*			
	Losses	*Sorties*	**Rate*	*Losses*	*Sorties*	**Rate*	*Losses*	*Sorties*	**Rate*	**Average Rate*
A-1	28	26,539	10.6	18	2,612	68.9	62	39,012	15.8	15.8
A-26	0	0	—	0	75	0	10	9,567	10.5	10.4
A-37	5	27,527	1.8	0	0	—	0	978	0	1.8
B-57	**15**	**15,488**	**9.7**	**5**	**3,089**	**16.2**	**11**	**13,225**	**8.3**	**9.7**
F-4	43	72,789	5.9	135	88,973	15.2	46	93,892	4.9	8.8
F-100	127	279,398	4.5	16	3,677	43.5	20	19,609	10.2	5.4
F-105	1	2,056	4.9	274	76,858	35.6	41	54,620	7.5	23.7
RB-57	**2**	**5,581**	**3.6**	**0**	**52**	**0**	**0**	**119**	**0**	**3.5**
RF-4	6	29,004	2.1	32	16,472	19.5	12	21,088	5.7	7.5
RF-101	2	14,154	1.4	27	9,726	27.8	3	8,701	3.4	9.8

*Loss rate per 10,000 sorties

APPENDIX 11
MARTIN B-57 UNIT ASSIGNMENTS

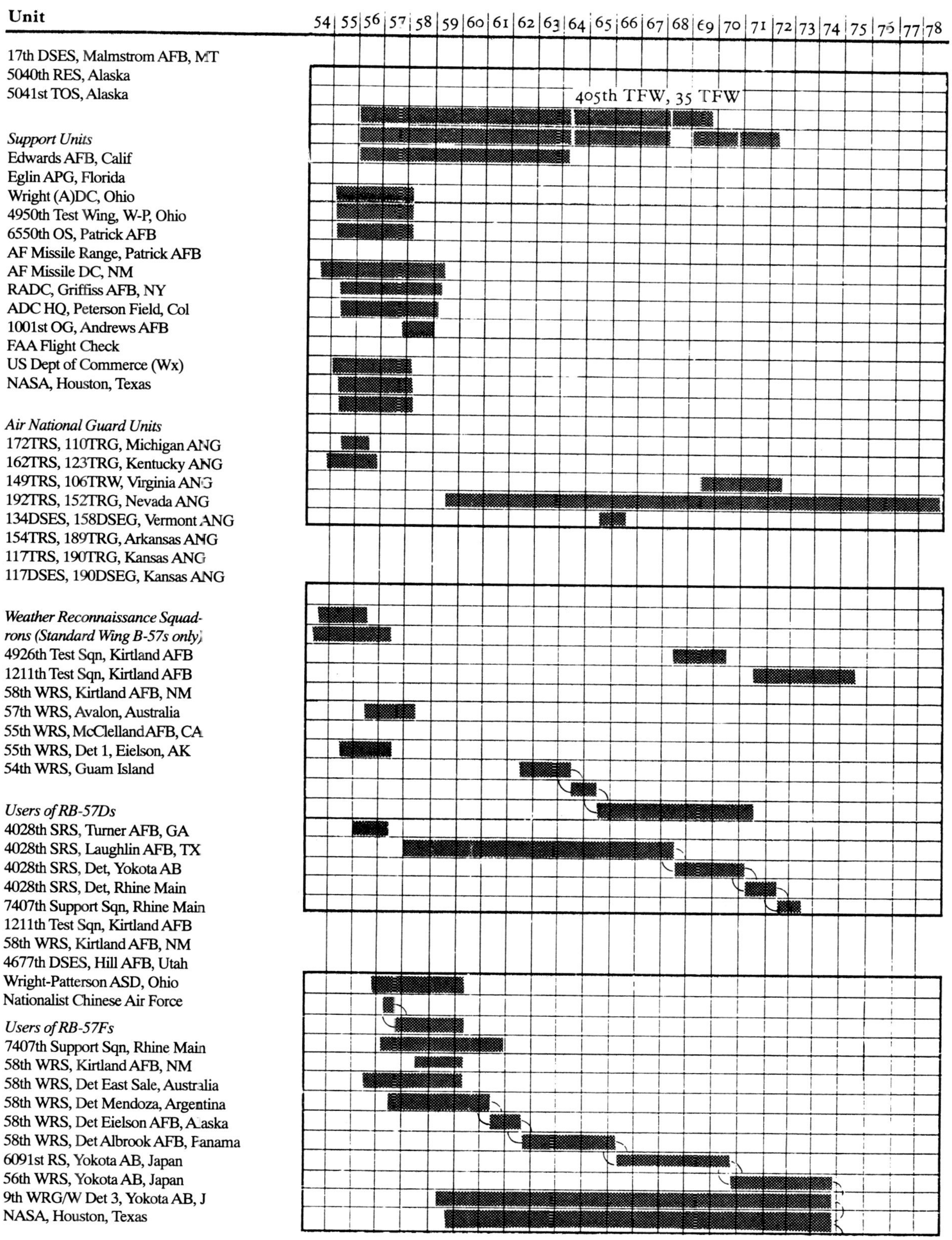

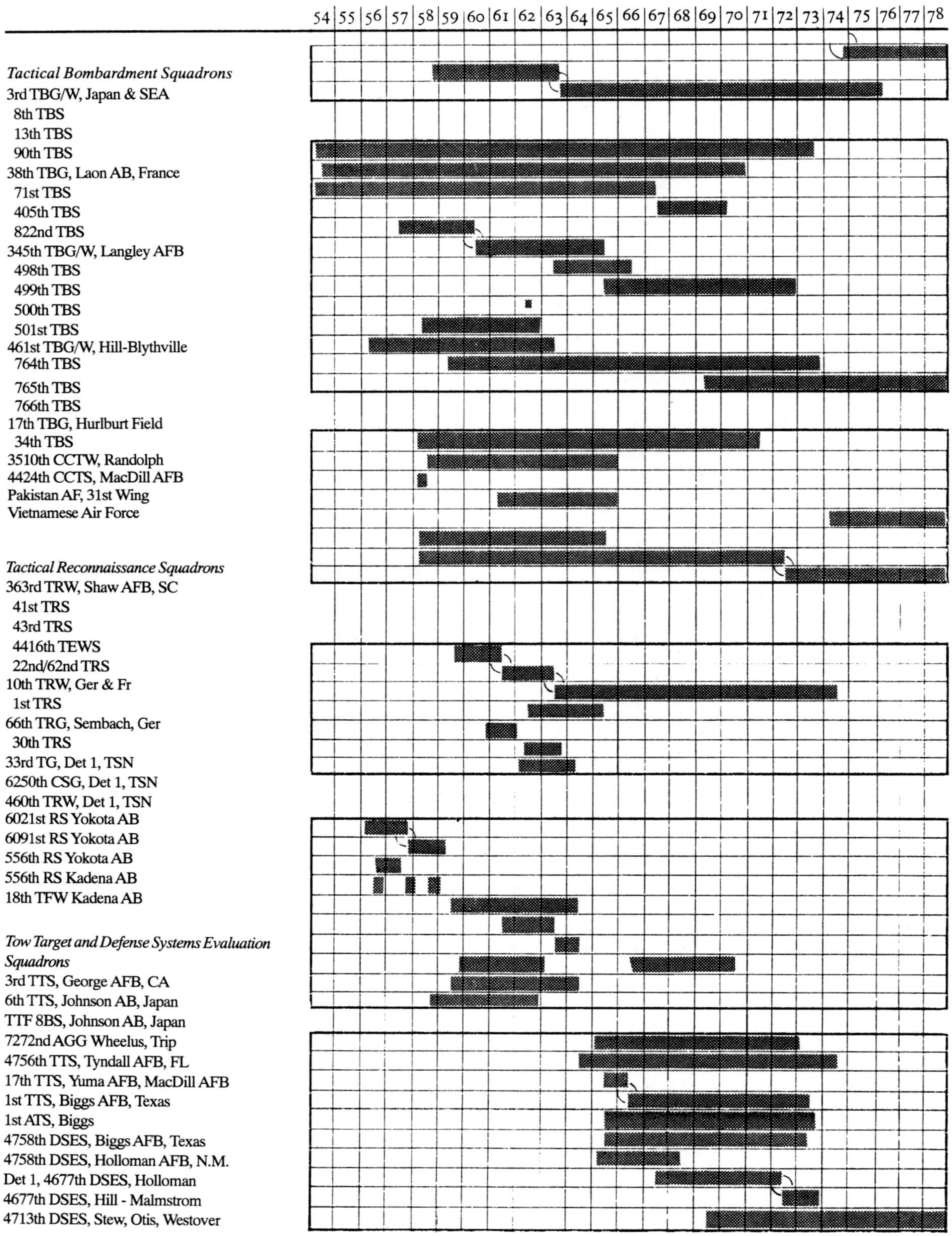
54 55 56 57 58 59 60 61 62 63 64 65 66 67 68 69 70 71 72 73 74 75 76 77 78
Tactical Bombardment Squadrons
3rd TBG/W, Japan & SEA
8th TBS
13th TBS
90th TBS
38th TBG, Laon AB, France
71st TBS
405th TBS
822nd TBS
345th TBG/W, Langley AFB
498th TBS
499th TBS
500th TBS
501st TBS
461st TBG/W, Hill-Blythville
764th TBS
765th TBS
766th TBS
17th TBG, Hurlburt Field
34th TBS
3510th CCTW, Randolph
4424th CCTS, MacDill AFB
Pakistan AF, 31st Wing
Vietnamese Air Force
Tactical Reconnaissance Squadrons
363rd TRW, Shaw AFB, SC
41st TRS
43rd TRS
4416th TEWS
22nd/62nd TRS
10th TRW, Ger & Fr
1st TRS
66th TRG, Sembach, Ger
30th TRS
33rd TG, Det 1, TSN
6250th CSG, Det 1, TSN
460th TRW, Det 1, TSN
6021st RS Yokota AB
6091st RS Yokota AB
556th RS Yokota AB
556th RS Kadena AB
18th TFW Kadena AB
Tow Target and Defense Systems Evaluation Squadrons
3rd TTS, George AFB, CA
6th TTS, Johnson AB, Japan
TTF 8BS, Johnson AB, Japan
7272nd AGG Wheelus, Trip
4756th TTS, Tyndall AFB, FL
17th TTS, Yuma AFB, MacDill AFB
1st TTS, Biggs AFB, Texas
1st ATS, Biggs
4758th DSES, Biggs AFB, Texas
4758th DSES, Holloman AFB, N.M.
Det 1, 4677th DSES, Holloman
4677th DSES, Hill - Malmstrom
4713th DSES, Stew, Otis, Westover

APPENDIX 12

MARTIN B-57 SURVIVORS

In museums and as gate-guards

Model Type	AF Serial	Hours Flown	Location
RB-57A	52-1426		Yankee AF, Ypsilanti, Mich.
RB-57A	52-1446		The GLM Aviation Museum, Baltimore
RB-57A	52-1456		MI ANG, 127 FW, Battle Creek, MI
RB-57A	52-1459		Florence Air & Missile Museum, SC
RB-57A	52-1467		Martin State Airport, Baltimore
RB-57A	52-1475		Robins AFB Museum, Macon, GA
RB-57A	52-1482		Lackland AFB, San Antonio, TX
RB-57A	52-1485		MI ANG, 127 FW, Battle Creek, MI
RB-57A	52-1488		New England A.M., Winsor Locks, CT
RB-57A	52-1492		Hill AFB Museum, Ogden, Utah.
B-57B	52-1499	7175	Air Force Museum, Dayton, OH
B-57B	52-1500	8433	Vermont ANG Base, Burlington VT
B-57B	52-1504	8712	Dyess AFB, Abilene, Texas
B-57E	52-1505	6037	Malmstrom AFB, Great Falls, MT
B-57B	52-1509	7605	Laughlin AFB, Del Rio, TX
B-57B	52-1516	9107	Eglin AFB, Valparaiso, FL
B-57B	52-1519	7737	March AFB, Riverside, CA
B-57B	52-1526	9605	Beale AFB, Marysville, CA
B-57B	52-1548	9502	Ellsworth AFB, Rapid City, SD
B-57B	52-1551		NASM, Smithsonian Inst., Wash.D.C.
B-57B	52-1576		Edwards AFB, California
B-57B	52-1584		Kalamazoo Avn. Hist'l Museum, MI
B-57C	53-3841		AK ANG, Little Rock, AK
RB-57D	53-3982		Pima Air Museum, Tucson, AZ
B-57E	55-4238		Puerto Rico Educ/Scient Foundation (Resides at Western Intl. Aircraft, Tucson, pending move)
B-57E	55-4244		SAC Museum, Bellevue, Nebraska
B-57E	55-4253		USAFM, location pending, at AMARC
B-57E	55-4274		Pima Air Museum, Tucson, AZ
B-57E	55-4276		San Clemente Museum, CA
B-57E	55-4279		Peterson AFB, Colorado Springs, CO
B-57E	55-4285		USAFM, location pending, at AMARC
B-57E	55-4293		Lowry AFB Museum, Denver, CO
RB-57F	63-13293		Robins AFB Museum, Macon, GA
RB-57F	63-13301		Pima Air Museum, Tucson, AZ
RB-57F	63-13302		USAFM, location pending, at AMARC
WB-57F	63-13501		Pima Air Museum, Tucson, AZ

In storage at AMARC awaiting disposition as of 28 December 1993:

EB-57B	52-1506	EB-57E	55-4239	WB-57F	63-13290
EB-57B	52-1511	EB-57E	55-4240	WB-57F	63-13291
EB-57B	52-1521	EB-57E	55-4241	WB-57F	63-13294
EB-57B	52-1545	EB-57E	55-4242	WB-57F	63-13295
EB-57B	52-1564	B-57E	55-4265	WB-57F	63-13300
B-57C	53-3831	EB-57E	55-4266	WB-57F	63-13301
B-57C	53-3840	EB-57E	55-4278	WB-57F	63-13302
B-57C	53-3856	EB-57E	55-4285		
		EB-57E	55-4295		
		EB-57E	55-4298		
		EB-57E	55-4300		

APPENDIX 13
MARTIN B-57 INDIVIDUAL ACTIVITY RECORDS

Location assignments found in Remarks section are from random notes and are not to be considered as complete aircraft histories. For full coverage of respective aircraft, record cards can be obtained from the Air Force History Center, Maxwell AFB, Alabama. Dates prefixed with a "c" (c3-69) implies circa, and not actual date of assignment or reassignment. Commas are used only when separating location assignments.

Key to Remarks section:

C-Loss: Combat loss followed by date of event.
Cx: Crashed followed by date of event.
ECM: Electronics Counter Measures
Eglin APG: Air Proving Ground, Eglin AFB, FL.
Flagstaff: 2 B-57Es at Andrews assigned to Off. AF Chief of Staff.
GLM: The Glenn L. Martin Co, builder of the B-57.
Gr-Ex: Ground explosion at Bien Hoa AB on 16 May 1965.
Heart Throb: A special trimming made to selected engines for the RB-57A for reaching higher altitudes.
MASDC: Now AMARC, Aerospace Maintenance and Regeneration Center, for storage, Davis Monthan AFB, Arizona.
Mo-Ak: Mortar Attack upon Bien Hoa AB, 1 November 1964.
NACA: National Advisory Committee for Aeronautics, later NASA.
Patrica Lynn: Spl. reconnaissance RB-57Es at Tan Son Nhut for testing equipment in combat.
RVN: Republic of Vietnam (South Vietnam).
SEA Conv: Identifies the 20 B-57s returned to Martin to be combat converted as SEA replacements.
Star Flight: 13 B-57C/Es at Andrews AFB for Gen Off adm. flying.
TAC: Tactical Air Command.
TM III: Tropic Moon III test program leading to B-57Gs.
USAFE: USAF in Europe, when no specific unit there is known.
W-ADC: Wright Aeronautical Development Center, Wright Patterson AFB, Ohio.
WRAMA: Warner-Robins Air Material Command, Robins AFB, Ga, depot for the B-57.
1/0 fat.: Number of fatalities. Only pilot in this example.
(Unit No.) See Appendix 11 for unit location and time period.

Special exceptions:
8/13BS-X: Aircraft assigned to one or the other squadron while stationed in Japan. Suffix letter is "tail letter."
8-BS or 13-BS: This format of the same units above for denoting SEA location.

(A) For scrap to Allied Aircraft Sales, Tucson, AZ 3-76.
(B) For scrap to Sun Valley Aviation, Phoenix, AZ 3-77.
(C) For scrap to NMT Terra Group, 6-79.
(D) For scrap to Airmet Corp., E. Tenn. St, Tucson, AZ 5-83
(E) For scrap to Allied Aircraft Sales, Tucson, AZ 8-84.
(F) In storage as of 28 December 1993 at MASDC (now AMARC).

Type	Ship No.	AF Serial	Final Comp.	Flown Away	Remarks
B-57A	1	52-1418	6-15-53	6-19-57	At GLM for performance, engine & a/c tests. 6-19-57 to NACA as 218.
B-57A	2	52-1419	7-29-53	12-3-53	Edwards AFB stab & cont tests, Kirtland 1-11-55, Wx Bureau N1005, Baker Aviation School, Miami International Airport.
B-57A	3	52-1420	8- 8-53		At GLM stab & cont tests, structural tests, Cx at GLM, 11-22-54. 2 fat.
B-57A	4	52-1421	11-16-53		At GLM arm. tests, Cx at Bel Air MD, (GLM) 12-9-53. 0 fat.
B-57A	5	52-1422	11-23-53	7-28-55	Eglin APG for cold hangar tests.
B-57A	6	52-1423	11-15-53	7-30-54	At GLM antenna testing, to W-ADC 7-30-54.
B-57A	7	52-1424	11-20-53	1-23-54	To Edwards 1-21-54, to Kirtland 10-13-54, Cx in Calif. 2-22-58.
B-57A	8	52-1425	11-30-53	1-18-54	To Edwards Shoran bombing tests, MASDC 73.
RB-57A	9	52-1426	11-16-53	1-18-55	At GLM for testing, MI ANG c68, Yankee AF, Ypsilanti, Mich c93.
RB-57A	10	52-1427	11-17-53	1-21-54	Heart Throb, Eglin APG 4-22-54, W-ADC 5-18-54, Yokota 6091 TRS, to RB-57F 53-13300.
RB-57A	11	52-1428	11-11-63	3-16-54	Edwards 6510 ABW 3-16-54 functional development, KS ANG c68, MASDC c73,(A).
RB-57A	12	52-1429	12-4-53	1-26-54	Heart Throb, W-ADC 7-2-54 functional development.
RB-57A	13	52-1430	12-11-53	3-22-54	Eglin APG 5-23-54, Cx at Eglin AFB 5-23-54. 2 fat.
RB-57A	14	52-1431	12-18-53	5-12-54	Heart Throb, Kirtland Special weapons tests 4925 TE, 5-12-54, Cx.
RB-57A	15	52-1432	12-23-53	3-22-54	Heart Throb, Eglin APG 3-22-54, to RB-57F 63-13301.
RB-57A	16	52-1433	12-31-53	3-22-53	Heart Throb, Eglin APG 3-22-54, RB-57F 63-13302.
RB-57A	17	52-1434	1-6-54	3-16-54	363 TRW 3-16-54, Eglin, Cx Keesler AFB MS 4-4-57. 1 fat.
RB-57A	18	52-1435	1-15-54	3-22-54	363 TRW 3-22-54, JRB-57A with APN-59 large nose c58 to end, NRB-57A Miami AACMF 9-25-68, RADC, MASDC 12-4-69,(A).
RB-57A	19	52-1436	1-21-54	3-16-54	363 TRW Shaw AFB 3-16-54.
RB-57A	20	52-1437	1-27-54	4-30-54	363 TRW 5-1-54, APN-59 project like '435 canx, MI ANG for ECM c3-66, 4758 DSES c9-68, MASDC c1-73, (A).
RB-57A	21	52-1438	1-28-54	3- 2-54	363 TRW 3-2-54, 801 ABG Lockbourn 8-11-54, FAA N96, Cx 6-7-60 control disconnect.
RB-57A	22	52-1439	1-29-54	3-30-54	Heart Throb, 363 TRW, 3-30-54, ECM conv. 3-66, 4758 DSES Holloman AFB c9-68, MASDC (A).

Type	Ship No.	AF Serial	Final Comp.	Flown Away	Remarks
RB-57A	23	52-1440	2-1-54	4-2-54	363 TRW, AR ANG, ECM conv. 3-66, 4758 DSES, MASDC c1-73, (A).
RB-57A	24	52-1441	2-4-54	4-14-54	363 TRW, Cx at Shaw AFB, 5-6-54. 2 fat.
RB-57A	25	52-1442	2-10-54	4- 5-54	Heart Throb, 363 TRW, ECM conv 3-66, 4758 DSES, MASDC c1-73, (A).
RB-57A	26	52-1443	2-16-54	4-13-54	363 TRW, Cx at Eglin AFB 8-6-58.
RB-57A	27	52-1444	2-17-54	4-19-54	363 TRW, Cx at Eglin AFB 8-6-58.
RB-57A	28	52-1445	2-22-54	4-23-54	363 TRW, Cx at Shaw AFB 10-6-5. 2 fat.
RB-57A	29	52-1446	2-25-54	4-30-54	363 TRW, 6091 RS Yokota c61, 56 WRS Yokota c63, KS ANG Topika c9-68, GLM Aviation Museum Baltimore c93.
RB-57A	30	52-1447	3-1-54	4-23-54	363 TRW 4-23-54, Eglin APG c54, FAA N97 7-16-57, USAF 8-12-63, ECM conv 3-66, 4758 DSES c68, MASDC (A).
RB-57A	31	52-1448	3-3-54	4-23-54	363 TRW, 6091 TRS c8-61, 56 WRS, ECM conv 3-66, 4758 DESE Holloman AFB.
RB-57A	32	52-1449	3-5-54	4-30-54	363 TRW, 6091 TRS Yokota c59, 56 WRS Yokota c63, KS ANG Topika c9-68.
RB-57A	33	52-1450	3-10-54	4-23-54	363 TRW, 6091 TRS c9-61, 57 WRS Guam, ECM conv 3-66, 4758 DESE c9-68, MASDC (A).
RB-57A	34	52-1451	3-17-54	4-30-54	363 TRW Shaw AFB.
RB-57A	35	52-1452	3-26-54	5-14-54	USAFE c54, KS ANG Topika c9-68, derelict at Forbes AFB c77.
RB-57A	36	52-1453	3-29-54	5-4-54	MI ANG Battle Creek c9-63, MASDC (A).
RB-57A	37	52-1454	4-1-54	5-17-54	USAFE, MI ANG c9-68, MASDC c1-73 (A).
RB-57A	38	52-1455	4-2-54	5-14-54	6091 TRS c61, 57 WRS Guam c63, KS ANG c68.
RB-57A	39	52-1456	4-6-54	5-13-54	USAFE 30 TRS, MI ANG c9-68, CAF Museum c73, MI ANG Battle Creek MI c93, on exhibit.
RB-57A	40	52-1457	4-8-54	6-11-54	363 TRW, KS ANG c9-68.
RB-57A	41	52-1458	4-12-54	5-13-54	MI ANG Battle Creek 5-20-54, GLM 7-12-54, dept Yokota 9-24-64, KS ANG c9-68.
RB-57A	42	52-1459	4-13-54	5-20-54	6091 TRS 'Switch Blade", (bogus s/n 52-1421 GX c63, 1618 GX c11-63, 1423 GX c64, 1423 GM, 1421 GX, MI ANG 9-68, Florence Air & Missile Museum N.C. c73.
RB-57A	43	52-1460	4-13-54	5-21-54	363 TRS, Sembach AB Germany, Cx Wheelus AB Tripoli 3-21-57. 0/1 fat.
RB-57A	44	52-1461	4-20-54	6-11-54	363 TRW, ECM conv 3-66, 4758 DSES c9-68, MASDC c1-73 (A).
RB-57A	45	52-1462	3-24-54	6-22-54	Heart Throb, 424 TBS (later 499 TBS) Langley AFB 6-22-54.
RB-57A	46	52-1463	4-26-54	6-11-54	363 TRW, KY ANG, MI ANG c9-68.
RB-57A	47	52-1464	4-27-54	6-21-54	Heart Throb, Eglin APG, ECM conv 3-66, 4758 DSES c9-68, MASDC (A).
RB-57A	48	52-1565	4-29-54	6-15-54	363 TRW, KS ANG Topika c9-68.
RB-57A	49	52-1466	5-3-54	6-22-54	Langley AFB 6-22-54, GLM 9-14-54, KS ANG Topeka c9-68.
RB-57A	50	52-1467	5-4-54	7-7-54	Langley AFB 7-7-54, GLM 10-11-54, 66 TRW, KS ANG c9-68, Aberdeen Prov. Gr., MD State Airport Baltimore Exhibit c93.
RB-57A	51	52-1468	5-6-54	11-10-54	66 TRW Simbach Ger 2-9-55, MI ANG, Cx Battle Creek 12-13-58.
RB-57A	52	52-1469	5-10-54	7-14-54	Langley AFB 7-14-54, GLM 9-23-54, Warner-Robins 1-11-55, ECM conv. 3-66, 4758 DSES c1-59, MASDC c1-73 (A).
RB-57A	53	52-1470	5-12-54	6-25-54	Langley AFB 6-25-54, GLM 9-22-54, Warner-Robins 1-11-55, Spangdahlem AB Germany Cx Germany 12-3-56. 2 fat.
RB-57A	54	52-1471	5-17-54	6-25-54	Langley AFB 6-25-54, GLM 10-1-54, Robins 11-3-54, MI ANG c9-68, Cx 9-69.
RB-57A	55	52-1472	5-21-54	1-20-55	Warner-Robins AFB Ga. (B-57 depot)
RB-57A	56	52-1473	5-21-54	11-10-54	66 TRW Simbach 2-9-55, 10 TRW Spangdahlem AB Ger 4-23-55, Cx Ger 4-23-55 with 52-1483 formation fly-by. 0 fat.
RB-57A	57	52-1474	5-26-54	11-12-54	W-Robins, MI ANG c9-68, MASDC c1-73 (A).
RB-57A	58	52-1475	5-27-54	11-24-54	363 TRW, USAFE 2-3-55, VA ANG 4-58 to 6-58, KS ANG Topika c9-68, Robins AFB Museum Macon GA c93.
RB-57A	59	52-1476	6-2-54	11-9-54	363 TRW, Cx at Shaw AFB 11-29-54. 2 fat.
RB-57A	60	52-1477	6-3-54	11-9-54	363 TRW Shaw AFB, NC.
RB-57A	61	52-1478	6-7-54	11-9-54	363 TRW, USAFE 2-3-55, MI ANG c9-68, MASDC (A).
RB-57A	62	52-1479	6-10-54	11-3-54	363 TRW, USAFE 1-25-55, KS ANG c9-68.
RB-57A	63	52-1480	6-8-54	11-9-54	363 TRW, USAFE 1-25-55, KS ANG c9-68.
RB-57A	64	52-1481	6-11-54	11-3-54	363 TRW, USAFE 1-26-55, MI ANG c9-68.
RB-57A	65	52-1482	6-15-54	11-22-54	Warner-Robins, USAFE 11-22-54, MI ANG c9-68, Lackland AFB Museum c73/c93.
RB-57A	66	52-1483	6-17-54	9-22-54	Warner-Robins, 10 TRW Spangdahlem AB Ger 10-20-54, Cx Ger 4-23-55 with 52-1473 formation fly-by.
RB-57A	67	52-1484	6-21-54	9-28-54	Warner-Robins, 10 TRW Spangdahlem AB Ger 10-20-54, 66 TRW Sembach 11-19-54, MI ANG Battle Creek c9-68, MASDC c73 (A).
RB-57A	68	52-1485	6-23-54	10-13-54	Warner-Robins, USAFE, MI ANG c9-68, MI ANG exhibit battle Creek MI c93.

Type	Ship No.	AF Serial	Final Comp.	Flown Away	Remarks
RB-57A	69	52-1486	6-25-54	10-14-54	Warner-Robins, 66 TRW Sembach 2-8-55, Cx Wiesbaden AB 12-23-55. 1 fat.
RB-57A	70	52-1487	6-28-54	10-20-54	Warner-Robins, 66 TRW Sembach AB 2-4-55, 10 TRW Spangdahlem AB 3-26-55, Cx Sembach 5-17-55. 2 fat.
RB-57A	71	52-1488	6-29-54	10-19-54	W-ADC, KS ANG c9-68, New England Air Museum Windsor Locks CT 6-72 c/93.
RB-57A	72	52-1489	6-30-54	11-12-54	Warner-Robins, USAFE, ECM conv, 4758 DESE Holloman AFB c9-68, MASDC c1-73 (A).
RB-57A	73	52-1490	7-2-54	10-27-54	363 TRW, W-Robins 1-20-55, USAFE, ECM conv 3-66, 4758 DESE c9-68, Cx Great Salt Lake 4-14-71.
RB-57A	74	52-1491	7-7-54	10-21-54	Warner-Robins, 66 TRW Sembach AB Ger 2-9-55, MI ANG c9-68, MASDC c12-72 (A).
RB-57A	75	52-1492	7-13-54	5-24-55	USAFE, had Flt Dir & Doppler system, Hill AFB Museum Ogden c93.
B-57B	76	52-1493	5-14-54	4-22-57	Bailed to GLM, Edwards armament tests.
B-57B	77	52-1494	7-21-54	9-3-54	Edwards Stab & Cont, Shoran, Cx Edwards 2-9-55. 1 fat.
B-57B	78	52-1495	7-26-54	8-31-54	GLM, Edwards 9-9-54, Cx Eds. 1-6-59
B-57B	79	52-1496	7-28-54	10-23-54	GLM, Spl Wpns Test Kirtland 10-23-54, MATS Wx, Cx 9-17-62 Australia.
B-57B	80	52-1497	7-29-54	8-31-54	Functional devel 3-54, Edwards, FAW 5-2-56 Bomarc nose.
B-57B	81	52-1498	8-16-54	9-29-54	Functional development 3-54, Edwards 9-29-54, SEA Conv, RVN, Clark, RVN, C-Loss 9-22-68. 0 fat.
B-57B	82	52-1499	8-18-54	2-24-56	GLM antenna test, bailed to Lear c5-64, SEA Conv c3-66, Clark 3-20-67, MASDC c10-69, ECM mod 8-72, KS ANG c6-77, VT ANG 3-79, USAF Museum 12-81 c92.
B-57B	83	52-1500	8-20-54	10-1-54	Eglin APG 10-2-54, 4677 DSES, KS ANG c6-77, VT ANG Burlington on exhibit.
B-57B	84	52-1501	8-25-54	10-1-54	Eglin APG 10-2-54.
B-57B	85	52-1502	8-26-54	1-4-55	Eglin APG, 4677 DSES c9-69, VT ANG c77, KS ANG, Cx mid-air.
B-57B	86	52-1503	8-27-54	10-13-54	Eglin APG, 4925 TG(A) Wx sampler, 4713 DSES c9-68, VT ANG 74, Cx Plattsburgh 1-14-80.
B-57B	87	52-1504	8-30-54	11-5-54	W-ADC, 4713 DSES 9-68, KS ANG 74, VT ANG 3-79, Dyess AFB Abilene exhibit.
B-57B	88	52-1505	8-31-54	10-7-54	Kirtland 1-26-55, 461 TBG Mobile-C 'B' 57, RADC 68-70, AMARC 4-20-70, Conv ECM 8-72, VT ANG 72-81, Malmstrom AFB 11-81 on exhibit c93.
B-57B	89	52-1506	9-2-54	10-13-54	GLM, Wx sampler Hill AFB, 4677 DSES, VT ANG c77, MASDC 12-81 (F).
B-57B	90	52-1507	9-3-54	1-20-55	KY ANG, Clark PI, RVN 1-65, MASDC 11-69, conv ECM 8-72, KS ANG 72, Cx 4-77.
B-57B	91	52-1508	9-8-54	1-12-55	KY ANG, Clark PI 1-65, C-loss 3-27-69. 0 fat.
B-57B	92	52-1509	9-8-54	1-21-55	345 TBG, 461 Mobile-C 'Q' 57, Patrick ETTRN 9-1-66, RADC 5-20-66, from JB-57B to B-57B 10-31-67, Miami AACMF 8-20-68, RADC c69, 4713 DSES c12-72, KS ANG c7-77, VT ANG 11-78, Laughlin AFB TX on exhibit 12-81 c93.
B-57B	93	52-1510	9-13-54	3-24-55	GLM, SEA Conv 65, Clark PI 12-11-65, RVN 12-11-65, C-Loss 9-14-67. 2 fat.
B-57B	94	52-1511	9-14-54	1-4-55	345 TBG, W-ADC, 4677 DSES c68, VT ANG c6-77, MASDC 12-81 (F).
B-57B	95	52-1512	9-16-54	1-21-55	345 TBG, 90 TBS c7-58, 13BS-P, Clark PI, RVN, Cx Dual flame-out 10-8-66. 2 fat.
B-57B	96	52-1513	9-17-54	1-21-55	345 TBG.
B-57B	97	52-1514	9-21-54	1-21-55	345 TBG, KY ANG Cx 9-11-58.
B-57B	98	52-1515	9-23-54	1-21-55	345 TBG.
B-57B	99	52-1516	9-24-54	1-27-55	345 TBG, SEA Conv 66, Clark PI, RVN 8-BS, MASDC 11-69, ECM conv, 8-72, VT ANG c77, Barksdale AFB LA exhibit 1981, Eglin AFB Valparaiso FL on exhibit c93.
B-57B	100	52-1517	9-27-54	2-9-55	461 TBG, Cx on delivery 2-9-55. 1/0 fat.
B-57B	101	52-1518	9-28-54	3-22-55	GLM, Eglin APG, NV ANG, Clark PI 5-23-65, RVN, TM III 5-27-67, Eglin c9-68, W-P ASD c6-69, MASDC c12-72 (E).
B-57B	102	52-1519	9-29-54	3-29-55	GLM, KY ANG, 8-BS 1-1-65, RVN, MASDC 11-69, ECM conv, VT ANG c7-72, March AFB CA on exhibit 12-81 c93.
B-57B	103	52-1520	10-1-54	3-25-55	GLM, W-ADC, 4677 DSES c9-68, 4677 DSES c12-72.
B-57B	104	52-1521	10-4-54	3-25-55	GLM, Edwards AFB c9-68, 4713 DSES c12-72, KS ANG c7-77, VT ANG 9-78, MASDC 12-81(F).
B-57B	105	52-1522	10-6-54	11-16-54	Randolph, 461 TBG Mobile-C 'Q' 57, Cx Little Rock 7-20-58.
B-57B	106	52-1523	10-7-54	11-16-54	Randolph, 461 TBG Mobile-C 'Q' 57, 13BS-C Clark PI 4-9-64, 8-BS 4-64, RVN, C-Loss 2-21-66. 0 fat.

Type	Ship No.	AF Serial	Final Comp.	Flown Away	Remarks
B-57B	107	52-1524	10-8-54	11-1-54	Randolph, Cx McChord AFB 11-19-58.
B-57B	108	52-1525	10-13-54	11-15-54	Randolph, 765 BS 461 TBG.
B-57B	109	52-1526	10-15-54	11-18-54	Randolph, 4713 DESE c9-68, KS ANG 7-77, VT ANG 8-78, Beale AFB CA on exhibit 12-81 c93.
B-57B	110	52-1527	10-18-54	12-1-54	Randolph, 461 TBG Mobile-C 'I' 57, to RB-57F 63-13289.
B-57B	111	52-1528	10-18-54	12-2-54	Randolph, Cx Randolph AFB TX 1-21-55. 2 fat.
B-57B	112	52-1529	10-20-54	12-1-54	Randolph, 461 TBG, Cx Webb AFB 12-4-55. 2 fat.
B-57B	113	52-1530	10-20-54	1-5-55	Randolph, NEV/KY ANG, SEA conv 56, RVN 5-65, C-Loss 4-7-66. 2 fat.
B-57B	114	52-1531	10-22-54	1-3-55	461 TBG, Cx Eniwetok Redwing 5-17-56. 0/1 fat.
B-57B	115	52-1532	10-25-54	1-5-55	Randolph, 90 BS c60, Clark PI 8-BS 4-64, Clark PI, RVN, VNAF, C-Loss date unk. 0 fat.
B-57B	116	52-1533	10-25-54	1-5-55	Randolph, Cx Eglin 6-22-58.
B-57B	117	52-1534	10-27-54	1-5-55	363 TRW, Randolph #33 c3-18-55, 461 TBG, Cx Tex/Mexico 11-29-55. 2 fat.
B-57B	118	52-1535	10-28-54	1-7-55	Randolph, Cx Randolph 6-22-55. 1 fat.
B-57B	119	52-1536	11-1-54	1-7-55	Randolph, 461 TBG Mobile-C 'M' 57, to RB-57F 63-13298.
B-57B	120	52-1537	10-29-54	1-5-55	Randolph.
B-57B	121	52-1538	11-2-54	12-5-55	Randolph, Cx Randolph 4-26-55. 0 fat.
B-57B	122	52-1539	11-8-54	1-5-55	Randolph, 461 TBG, TM-76 MACE nose Sembach Ger 61, to RB-57F 63-13288.
B-57B	123	52-1540	11-9-54	1-5-55	Randolph, 461 TBG Mobile-C 'H' 57, Patrick AFB with photo test pattern 1-22-60.
B-57B	124	52-1541	11-12-54	1-7-55	Randolph, 13BS-M, Clark PI, RVN C-Loss 9-19-66 2 fat.
B-57B	125	52-1542	11-15-54	1-7-55	Randolph.
B-57B	126	52-1543	11-16-54	1-7-55	Randolph, 8BS-L, Cx Yokota 2-4-61.
B-57B	127	52-1544	11-17-54	1-7-55	Randolph, 461 TBG Mobile-C 'G' 57, NV ANG, Clark PI 5-23-65, RVN, C-Loss 9-6-65. 0 fat.
B-57B	128	52-1545	11-19-54	1-12-55	345 TBG, KS ANG, 8-BS Clark PI 5-65, MASDC 11-69, ECM Conv 8-72, VT ANG, MASDC 12-81 (F).
B-57B	129	52-1546	11-22-54	1-18-55	Randolph, 345 TBG, Cx Langley 3-19-56. 2 fat.
B-57B	130	52-1547	11-26-54	1-7-55	Randolph.
B-57B	131	52-1548	11-29-54	1-12-55	Randolph, 4677 DSES c9-68, KS ANG c77, VT ANG 3-79, Ellsworth AFB SD Museum.
B-57B	132	52-1549	12-6-54	1-12-55	3750 Technical Training Wing.
B-57B	133	52-1550	12-8-54	2-9-55	Randolph, TAC, SEA Conv 65, Clark PI 12-20-65, RVN, C-Loss 3-5-66. 0 fat.
B-57B	134	52-1551	12-16-54	1-21-55	345 TBG, KY ANG, Clark PI 1-65, 8-BS lost tail-pipe 3-10-69, rt wg from 552, MASDC 11-69, ECM conv Birmingham 9-72, 4713 DSES 11-73, KS ANG c77, VT ANG 9-78, NASM Dulles storage 10-20-81.
B-57B	135	52-1552	11-19-54	1-21-55	345 TBG, Redwing 5-56, MACE nose, MASDC prior 69.
B-57B	136	52-1553	12-20-54	1-26-55	345 TBG, 461 TBW, AMARC.
B-57B	137	52-1554	12-23-54	2-9-55	W-Robins AMC, TAC, Cx Hill 7-28-55. 1 fat.
B-57B	138	52-1555	12-26-54	3-25-55	GLM, 90BS-N c7-58, 8/13-BS Clark PI, Mo-Ak 11-1-6.
B-57B	139	52-1556	12-15-54	2-9-55	461 TBG, Ret GLM for repair 12-16-56.
B-57B	140	52-1557	12-29-54	2-9-55	461 TBG, Mobile-C 'M' 57, SEA Conv 3-66, Clark PI, RVN, C-Loss 1-21-67. 2 fat.
B-57B	141	52-1558	12-30-54	3-25-55	461 TBG Mobile-C 'A' 57, 4713 DSES.
B-57B	142	52-1559	1-4-54	3-29-55	TAC, to RB-57F 63-13286.
B-57B	143	52-1560	1-6-55	3-29-55	USAFE.
B-57B	144	52-1561	1-13-55	3-26-55	USAFE, KY ANG, Clark PI 1-23-65, RVN, C-Loss 1-13-69. 2 fat.
B-57B	145	52-1562	1-14-55	3-25-55	TAC, TM-76 MACE nose, to RB-57F 63-13290.
B-57B	146	52-1563	1-17-55	3-28-55	461 TBG, Cx Langley AFB 9-8-55. 0 fat.
B-57B	147	52-1564	1-19-55	3-30-55	USAFE, AFLC Hill AFB 9-69, 4677 12-72, VT ANG c7-77, MASDC 12-81 (F).
B-57B	148	52-1565	1-30-55	3-28-55	38 TBW, NV/KY ANG, Clark PI 5-7-65, 8-BS C-Loss 12-14-65. 1/0 fat.
B-57B	149	52-1566	2-1-55	3-29-55	38 TBG, Cx 4-2-57. 2 fat.
B-57B	150	52-1567	2-2-55	3-30-55	38 TBG, NV/KY ANG, 8-BS 5-65, Cx dual flame-out Phan Rang 3-15-69. 0 fat.
B-57B	151	52-1568	2-5-55	3-29-55	38 TBG, 8BS-Z, Clark PI 4-9-64, RVN, Gr-Ex 5-16-65.
B-57B	152	52-1569	2-8-54	2-3-56	38 TBG, Redwing Cx 1-9-59.
B-57B	153	52-1570	2-11-55	3-31-55	38 TBG, NV ANG, Clark PI 5-23-65, RVN, C-Loss 10-5-68. 0 fat.
B-57B	154	52-1571	2-14-55	3-31-55	38 TBG, SEA Conv 3-66, 8-BS Clark PI 1-21-66, MASDC 11-69, ECM conv Birmingham 8-72, VT ANG, Cx 3-76.
B-57B	155	52-1572	2-15-55	4-1-55	38 TBG, Cx Edwards 12-24-58.
B-57B	156	52-1573	2-17-55	4-11-55	38 TBG, to RB-57F 63-13299.
B-57B	157	52-1574	2-22-55	4-15-55	38 TBG 71 BS, to RB-57F 63-13291.

Type	Ship No.	AF Serial	Final Comp.	Flown Away	Remarks
B-57B	158	52-1575	2-24-55	4-12-55	38 TBG, KY ANG, Clark PI 1-23-65, RVN, 13-BS C-Loss 2-10-66. 2 fat.
B-57B	166	52-1576	11-16-54	1-21-56	Bailed to GLM, 1st with 20mm, bailed to NACA, Edwards AFB CA on exhibit c93.
B-57B	167	52-1577	3-7-55	5-26-55	38 TBG 822 BS.
B-57B	168	52-1578	1-10-55	5-16-55	38 TBG, SEA Conv 3-66, 8-BS Clark PI, B-57G 5-69, Ubon 9-70, KS ANG 12-72, MASDC 74 (B).
B-57B	169	52-1579	1-17-55	5-16-55	38 TBG.
B-57B	170	52-1580	1-24-55	5-16-55	38 TBG, SEA Conv, Clark PI 66, RVN, TM III Eglin, B-57G 5-69, Ubon 9-70, KS ANG 12-72, MASDC 74 (B).
B-57B	171	52-1581	1-27-55	5-19-55	38 TBG, Terrain following Radar (Cornell) 10-1-60/1-1-61, W-P AFLC 9-68, MASDC 8-72 (A).
B-57B	172	52-1582	2-7-55	5-31-55	38 TBG, SEA Conv., 8-BS Clark PI, B-57G 5-69, Ubon 9-70, KS ANG 4-72, MASDC 74 (B).
B-57B	173	52-1583	2-16-55	5-19-55	38 TBG, to RB-57F 63-13293
B-57B	174	52-1584	2-23-55	5-20-55	38 TBG, W-ASD 66, MASDC 7-70, GNB-57B 3345 Ma Sup Gr c12-72, Chanute AFB Museum, Kalamazoo Aviation Museum c93.
B-57B	175	52-1585	2-28-55	6-2-55	38 TBG.
B-57B	176	52-1586	3-2-55	5-23-55	38 TBG, SEA Conv, RVN 7-30-66, C-Loss 4-4-68. 0 fat.
B-57B	177	52-1587	3-3-55	5-24-55	38 TBG, Cx Sembach Ger 3-26-57. 2 fat.
B-57B	178	52-1588	3-10-55	5-31-55	38 TBG, SEA Conv., 8-BS Clark PI 5-22-66, B-57G 5-69, Ubon 9-70, KS ANG 5-72, MASDC 74 (B).
B-57B	179	52-1589	3-11-55	6-1-55	Unk, KY ANG, RVN 1-1-65, C-Loss 6-30-65. 0 fat.
B-57B	180	52-1590	3-14-55	6-2-55	38 TBG, SEA Conv, Clark PI 1-11-66, RVN, C-Loss 12-8-66. 0 fat.
B-57B	181	52-1591	3-17-55	6-3-55	38 TBG, Cx Laon, France 12-11-56. 2 fat.
B-57B	182	52-1592	3-18-55	6-15-55	38 TBG, NV ANG, RVN 4-65, C-Loss 3-25-68. 2 fat.
B-57B	183	52-1593	3-21-55	6-6-55	38 TBG.
B-57B	184	52-1594	3-23-55	5-26-55	38 TBG, to RB-57F 63-13292.
B-57C	159	55-3825	11-18-54	4-22-55	Cx Randolph 6-11-55. 2 fat.
B-57C	160	53-3826	1-10-55	2-9-55	Randolph 2-8-55.
B-57C	161	53-3827	1-25-55	4-19-55	Unk, SEA Conv, 8-BS, Clark PI, MacDill 12-1 68, MASDC c2-70/73 (A).
B-57C	162	53-3828	2-2-55	4-19-55	Unk, KY ANG, 8-BS Clark PI 1-1-65, MASDC 11-69 (A).
B-57C	163	53-3829	2-14-55	4-27-55	461 TBG, Cx Wendover 3-26-57. 2 fat.
B-57C	164	53-3830	2-21-55	5- 7-55	Unk, Cx Buena Vista VA 3-20-58.
B-57C	165	53-3831	3-1-55	6-10-55	Ger, NV ANG, SEA Conv c65, 4677 DSES c9-68, 4677 c12-72, VT ANG 4-74, MASDC 12-81 (F).
B-57C	187	53-3832	3-9-55	6-15-55	TAC, Star Flight Andrews 58/62, Wx sampler 62, 58 WRS Kirtland.
B-57C	190	53-3833	3-16-55	6-4-55	3BW-H, 8/13-BS Clark PI 4-18-64, RVN, C-Loss 4-17-66. 0 fat.
B-57C	192	53-3834	3-21-55	6-22-55	USAFE, Pakistan AF 6-59, Cx 3-2-60
B-57C	194	53-3835	3-31-55	6-14-55	TAC, Cx 8-29-61.
B-57C	197	53-3836	4-28-55	7-8-55	3BW-E, 13BS-E, Clark PI 4-9-64, Cx training msn Clark AB 6-4-64. 2 fat.
B-57C	201	53-3837	4-15-55	6-17-55	90BS-C, Star Flight Andrews 1958-62, Wx Conv 62, 58 WRS.
B-57C	205	53-3838	4-20-55	6-20-55	TAC, SEA Conv 65, Clark PI 12-22-65, 13-BS Cx Cubi PI 8-1-66. 0 fat.
B-57C	209	53-3839	4-27-55	6-23-55	USAFE, Star Flight Andrews c61, Wx Conv 62, 58 WRS c68, Cx 4-15-70 sgl-eng tng.
B-57C	213	53-3840	5-3-55	6-29-55	8BS-E 1-9-56, 8-BS-B Clark PI 4-9-64, MacDill 12-68, KS ANG 12-72, VT ANG 5-79, MASDC 12-81 (F).
B-57C	217	53-3841	5-11-55	7-21-55	38 TBG 822 BS, KS ANG c9-68, Little Rock pylon exhibit 78 c93.
B-57C	221	53-3842	5-6-55	10-12-55	Unk, Star Flight Andrews 58-62, Wx conv 62, 58 WRS c-9-68, mid-air with Cessna 9-25-71, flyable, AMARC 12-72 (C).
B-57C	225	53-3843	5-12-55	9-26-55	W-Robins AMC,Cx Robins AFB 1-4-56. 0 fat.
B-57C	229	53-3844	5-26-55	8-17-56	Conv to tow target 8-6-58.
B-57C	233	53-3845	5-24-55	5-31-56	Conv to tow target.
B-57C	237	53-3846	6-8-55	9-22-55	345 TBG, Pakistan AF 1959.
B-57C	241	53-3847	6-7-55	9-30-55	GLM, Cx Yuma 10-2-55. 1 fat.
B-57C	245	53-3848	6-10-55	9-20-55	Unk, Laughlin AFB TX, Cx Andrews AFB 9-12-57. 2 fat.
B-57C	249	53-3849	6-20-55	9-30-55	8BS-K, 8-BS Clark PI 4-18-64, wing chg in US 12-65, Clark PI 5-66, MacDill 12-68, MASDC 12-72 (A).
B-57C	253	53-3850	6-30-55	9-21-55	Unk, Star Flight Andrews 58-62, Wx conv 62, 58 WRS c68, MASDC 12-72 (C).
B-57C	257	53-3851	7-7-55	9-23-55	Unk, Star Flight Andrews 58-62, Wx conv 62, 58 WRS c68, MASDC 12-72 (C).
B-57C	261	53-3852	7-12-55	9-23-55	38 TBG. Cx in France 3-14-57. 2 fat.
B-57C	265	53-3853	7-14-55	10-18-55	Unk, Cx at Andrews 1-25-58.
B-57C	269	53-3854	7-22-55	10-3-55	13BS-H 6-57, 90BS to 2-18-62, RADC 66-72, MASDC 7-20-72 (A).

Type	Ship No.	AF Serial	Final Comp.	Flown Away	Remarks
B-57C	273	53-3855	7-27-55	10- 4-55	90BS-T, Cx 5-7-58.
B-57C	277 5	3-3856	8-16-55	9-29-55	3BW-V, 3-BS Clark PI, MacDill 12-68, 4677 DSES c72, VT ANG 4-74, MASDC 12-81 (F).
B-57C	281	53-3857	8-22-55	10- 3-55	Unk, 58 WRS c9-68, MASDC (C).
B-57C	285	53-3858	8-25-55	2-13-56	345 TBG, Cx at Langley 5-8-56. 2 fat.
B-57B	185	53-3859	3-24-55	6-6-55	USAFE, SEA Conv 66, Clark PI, RVN, C-Loss 4-22-67. 2 fat.
B-57B	186	53-3860	3-28-55	6-7-55	3BG, NV/KY ANG, Clark PI 5-23-65, RVN TM II, B-57G 69, Ubon 9-70, KS ANG 5-72, MASDC 74 (E).
B-57B	188	53-3861	3-30-55	6-10-55	3 BW.
B-57B	189	53-3862	4-1-55	6-13-55	3 BW, Unk, SEA Conv, RVN, C-Loss 6-3-67. 2 fat.
B-57B	191	53-3863	4-4-55	6-10-55	3 BW.
B-57B	193	53-3864	4-5-55	6-13-55	3 BW, to RB-57F 63-13287.
B-57B	195	53-3865	4-8-55	6-14-55	3 BW, Unk, SEA Conv, 8-BS 3-15-66, B-57G 5-69, Ubon 9-70, KS ANG 72, MASDC 74(E).
B-57B	196	53-3866	4-11-55	6-16-55	3 BW, Unk, Cx WRAMA 11-9-55.
B-57B	198	53-3867	4-13-55	6-28-55	90 BS-C, 8/13BS Clark PI, RVN, Gr-Ex
B-57B	199	53-3868	4-18-55	6-20-55	3 BW, Cx Japan 7-19-57. 0 fat.
B-57B	200	53-3869	4-19-55	6-16-55	3 BW.
B-57B	202	53-3870	4-25-55	6-21-55	8BS-D, Clark PI 4-18-64, TSN C-loss 8-6-64. 2 fat.
B-57B	203	53-3871	4-26-55	6-23-55	13BS-I, Clark PI, Bien Hoa Gr-Ex 5-16-65.
B-57B	204	53-3872	5-2-55	6-29-55	To Japan Cx at Luke 12-16-55. 0 fat.
B-57B	206	53-3873	5-10-55	6-27-55	90BS-D, Clark PI 4-24-64, Gr-Ex 5-16-65.
B-57B	207	53-3874	5-12-55	7-12-55	Cx GLM 7-12-55. 0 fat.
B-57B	208	53-3875	5-16-55	7-8-55	3 BW 1-9-56, Cx Japan 9-13-56. 0 fat.
B-57B	210	53-3876	5-13-55	7-20-55	13BS-A, Clark PI 4-10-64, DaNang 13-BS C-Loss 1-11-66. 2 fat.
B-57B	211	53-3877	5-25-55	9-22-55	90BS-G, 8-BS Clark PI 4-17-64, B-57G 69, MacDill, KS ANG 5-72, MASDC 74 (E).
B-57B	212	53-3878	5-24-55	9-28-55	13BS-W, 8-BS Clark PI 4-17-64, B-57G 69, MacDill, KS ANG 5-72, MASDC 74.
B-57B	214	53-3879	5-27-55	10-5-55	8BS-R, 3-BS Clark PI 4-18-64, RVN, MASDC 11-69 (A).
B-57B	215	53-3880	5-31-55	9-30-55	8BS-T, 8-BS Clark 4-9-64, C-Loss 4-6-65. 2 fat.
B-57B	216	53-3881	6-3-55	10-7-55	8BS-I, Cx Japan 6-23-61.
B-57B	218	53-3882	6-9-55	10-7-55	8BS-P/V, 8-BS Clark PI, C-Loss 6-8-65. 0 fat.
B-57B	219	53-3883	6-10-55	10-12-55	3 BW, Cx Japan 11-7-57. 2 fat.
B-57B	220	53-3884	6-16-55	10-17-55	90BS-F, Clark PI, Cx Bien Hoa 8-6-64. 0 fat.
B-57B	222	53-3885	6-22-55	3-31-56	461 TBG Mobile-C 'P' 57, Pakis. AF 1959.
B-57B	223	53-3886	6-23-55	11-14-55	13BS-J, 8-BS Clark PI 4-17-64, RVN, B-57G 69, Ubon 9-70, KS ANG 5-72, Cx 72.
B-57B	224	53-3887	6-28-55	12-10-55	13BS-Q, Cx Japan 7-25-58.
B-57B	226	53-3888	7-1-55	10-14-55	90BS-E, Clark PI 4-17-64, RVN, C-Loss 10-6-66. First jet to bomb in VN after Geneva Conv rules broken. 0/1 fat.
B-57B	227	53-3889	7-6-55	3-8-56	90BS-M, 8-BS Clark PI 4-10-64, RVN, B-57G 69, Ubon 9-70, KS ANG 5-72, MASDC 74 (E).
B-57B	228	53-3890	7- 8-55	5- 3-56	13BS-C, 13-BS Clark PI 4-10-64, RVN, 8-BS C-Loss 3-11-65. 2 fat.
B-57B	230	53-3891	7-13-55	3-31-56	345 TBG, Pakistan AF 1959.
B-57B	231	53-3892	7-19-55	2-29-56	8BS-C, Clark PI 4-9-64, Mo-Ak 11-1-64.
B-57B	232	53-3893	7-21-55	10-17-55	13BS-N, Clark PI 4-17-64, RVN, Gr-Ex 5-16-65.
B-57B	234	53-3894	7-22-55	10-10-55	8BS-S, Clark PI 4-17-64, Mo-Ak 11-1-64.
B-57B	235	53-3895	7-25-55	12-1-55	8BS-U/N, Clark PI 5-14-64, RVN C-loss 6-29-65. 2 fat.
B-57B	236	53-3896	7-29-55	11-8-55	3 BG, Cx Japan 10-25-56. 0 fat.
B-57B	238	53-3897	8-19-55	11-2-55	8BS-B, RB-57F 63-13296.
B-57B	239	53-3898	8-19-55	11-3-55	8BS-H, Clark PI 4-9-64, RVN, B-57G 69, Ubon 9-70, KS ANG 5-72, MASDC 74 (B).
B-57B	240	53-3899	8-23-55	7-18-56	Unk.
B-57B	242	53-3900	8-26-55	11-18-55	13BS-M, RB-57F 63-13297.
B-57B	243	53-3901	8-30-55	11-9-55	3 BG, Cx Japan 10-20-57. 0/1 fat.
B-57B	244	53-3902	9-2-55	11-11-55	13BS-K.
B-57B	246	53-3903	9-6-55	12-6-55	90BS-J, Clark PI 4-9-64, RVN, C-Loss 1-16-66. 0 fat.
B-57B	247	53-3904	9- 9-55	2-22-56	13BS-B, Clark PI 4-10-64, RVN, Gr-Ex 5-16-65.
B-57B	248	53-3905	9-13-55	2-21-56	90BS-X, Clark PI 4-21-64, RVN, B-57G 7-68, Cx GLM 12-16-69.
B-57B	250	53-3906	9-16-55	2-22-56	90BS-Y, 8 BS Clark PI 4-17-64, RVN, B-57G 69, MacDill, Pave Gat w/ turret 2-4-71/7-31-71, KS ANG 72, MASDC 74 (E).
B-57B	251	53-3907	9-21-55	12-5-55	90BS-H, Cx Japan 6-3-59.
B-57B	252	53-3908	9-26-55	2-21-56	90BS-I, Clark PI 4-10-64, RVN, C-Loss 6-9-67. ? fat.
B-57B	254	53-3909	9-29-55	3-20-56	345 BG, Cx Langley 5-26-58.
B-57B	255	53-3910	9-30-55	2-14-56	13BS-F, Clark PI 4-29-64, RVN, C-Loss 6-21-65. 2 fat.
B-57B	256	53-3911	10-3-55	11-15-55	8BS-M.

Type	Ship No.	AF Serial	Final Comp.	Flown Away	Remarks
B-57B	258	53-3912	10-5-55	5-18-56	Unk.
B-57B	259	53-3913	10-10-55	11-17-55	90BS-L, Clark PI 4-17-64, RVN, Gr-Ex 5-16-65.
B-57B	260	53-3914	10-11-55	12-10-55	90BS-S, Clark PI 4-10-64, RVN, Mo-Ak 11-1-64.
B-57B	262	53-3915	10-14-55	11-22-55	8BS-J, Clark PI 4-17-64, Gr-Ex 5-16-65.
B-57B	263	53-3916	10-18-55	2-28-56	3 BG, Cx ferry flight 5-12-57. 0 fat.
B-57B	264	53-3917	10-19-55	6-25-56	Returned for Repair Fort Worth 7-19-56.
B-57B	266	53-3918	10-21-55	1-12-56	13BS-D, 1-11-64 to US, RB-57F 63-13295.
B-57B	267	53-3919	10-26-55	2-20-56	90BS-Z, Clark PI 4-17-64, RVN, C-Loss 8-6-65. 0 fat.
B-57B	268	53-3920	10-27-55	7-19-56	NV ANG, RVN 4-65, 8-BS C-Loss 10-20-65. 0 fat.
B-57B	270	53-3921	10-31-55	7-19-56	Cx Stewart 10-1-59.
B-57B	271	53-3922	11-3-55	3-5-56	Unk.
B-57B	272	53-3923	11-7-55	7-19-56	345 GB, Cx A.P. Hill 5-14-57. 2 fat.
B-57B	274	53-3924	11-9-55	10-1-56	13BS-X, Clark PI 4-20-64, RVN, Mo-Ex 11-1-64.
B-57B	275	53-3925	11-11-55	1-6-56	13BS-G, Clark PI 5-11-64, Cx 4-14-65. 0 fat.
B-57B	276	53-3926	11-15-55	2-20-56	8BS-A, Clark PI 4-18-64, RVN, Cx 4-14-65, 0 fat.
B-57B	278	53-3927	11-17-55	7-18-56	345 BG, Cx 12-14-57 (lost bomb bay door).0 fat.
B-57B	279	53-3928	11-21-55	9-5-56	Unk, SEA Conv c3-66, Clark PI 4-17-66, RVN, VNAF, B-57G 7-15-68, MacDill c69, KS ANG 72, MASDC 74 (B).
B-57B	280	53-3929	11-22-55	7-21-56	Unk, NV/KY ANG, 8-BS 5-65, VNAF, B-57G 69, Ubon 9-70 (rt wg from 282), KS ANG 72, MASDC 74 (B).
B-57B	282	53-3930	11-29-55	7-21-56	461 BG Mobile-C 'Q',13BS-Q, Clark PI 4-10-64, RVN, Gr-Ex 5-16-65.
B-57B	283	53-3931	12-5-55	3-2-56	8BS-W/Q, 8 BS, Clark PI 4-9-64, RVN, B-57G 69, Ubon 9-70, Cx with O-2A at night over Laos 12-12-70. 0 fat.
B-57B	284	53-3932	12-9-55	3-2-56	13BS-L, Cx Japan 8-24-59
B-57B	286	53-3933	12-14-55	4-9-56	Unk.
B-57B	287	53-3934	12-16-55	3-21-56	461 BG Mobile-C 'J' 57.
B-57B	288	53-3935	12-21-55	3-30-56	8 BS-F, to US 1-11-64, RB-57F 63-13294.
B-57C	289	53-3936	8-31-55	12-4-55	345 TBG, Cx Langley 4-30-57 mid-air. 0/1 fat.
B-57B	290	53-3937	12-28-55	3-30-56	345 BG, 8/13-T, Clark PI 4-10-64, RVN, Gr-Ex 5-16-65.
B-57B	291	53-3938	12-30-55	7-19-56	Unk, Pakistan AF 1959.
B-57B	292	53-3939	1-5-56	3-31-56	Unk, Palmdale c12-61, Pakistan AF 1959.
B-57C	293	53-3940	9-8-55	1-3-57	Unk, KY ANG, 8-BS Clark PI 1-1-65, MacDill 12-68, KY ANG c12-72,MASDC (C).
B-57B	294	53-3941	1-6-56	4-9-56	Unk, Pakistan AF 1959.
B-57B	295	53-3942	1-11-56	3-31-56	Unk, Pakistan AF 1959.
B-57B	296	53-3943	12-16-55	3-31-56	461 BG Mobile-C 'L' 57, Pakistan AF 1959.
B-57C	297	53-3944	9-15-55	3-30-56	Unk, Star Flight Andrews 58-62, Wx conv, 58 WRS c9-68, MASDC c12-72.
B-57B	298	53-3945	1-20-56	3-31-56	Unk, Pakistan AF 1959.
B-57B	299	53-3946	1-23-56	4-14-56	Unk, Pakistan AF 1959.
B-57B	300	53-3947	1-26-56	4-25-56	Unk, Pakistan AF 1959.
B-57C	301	53-3948	9-27-55	8-14-56	Turner AFB, KS ANG c12-61. (was sched for Pakistan AF)
B-57B	302	53-3949	1-31-56	4-20-56	Unk, Pakistan AF 1959.
B-57B	303	53-3950	2-1-56	4-19-56	Unk, Pakistan AF 1959.
B-57B	304	53-3951	2-7-56	4-20-56	461 BG Mobile-C 'K' 57, Pakistan AF 1959.
B-57B	305	53-3952	2-8-56	4-19-56	Unk, Pakistan AF 1959.
B-57B	306	53-3953	2-14-56	5-8-56	345 BG, Cx 3-21-60.
B-57B	307	53-3954	2-23-56	7-21-56	461 Mobile-C 'W', Pakistan AF 1959.
B-57B	308	53-3955	2-24-56	5-8-56	Unk, Pakistan AF 1959.
B-57B	309	53-3956	3-2-56	4-18-56	Unk, Pakistan AF 1959.
B-57B	310	53-3957	3-12-56	4-25-56	Unk, Pakistan AF 1959
B-57B	311	53-3958	3-13-56	5-8-56	Unk, Pakistan AF 1959.
B-57B	312	53-3959	3-16-56	7-21-56	Unk, Pakistan AF 1959.
B-57B	313	53-3960	3-26-56	12-7-56	345 BG, Pakistan AF 1959.
B-57B	314	53-3961		10-1-56	Unk, Pakistan AF 1959.
B-57B	315	53-3962	4-31-56	6-5-56	345 BG, Cx Langley 6-25-58
RB-57D	014	53-3963	10-25-56	6-13-57	4028 SRS, MASDC.
RB-57D	015	53-3964	10-18-56	6-6-57	W-ADC, 4677 DSES c9-68, MASDC c72 (A).
RB-57D	016	53-3965	10-11-56	3-8-57	4028 SRS, MASDC, 4677 DSES c9-68, MASDC c72 (A).
RB-57D	017	53-3966	11-2-56	7-15-57	4028 SRS, 4677 DSES c9-68, MASDC c72.
RB-57D	018	53-3967	12-10-56	2-10-57	4028 SRS, MASDC, 4677 DSES c9-68, MASDC c72 (A).
RB-57D	019	53-3968	12-19-56	3-13-57	4028 SRS, MASDC, 4677 DSES c9-68, MASDC c72.
RB-57D	020	53-3969	2-9-57	3-30-57	4028 SRS, MASDC, 4677 DSES c9-68, MASDC c72 (A).
RB-57D	007	53-3970	3-2-56	8-26-56	4028 SRS, MASDC, RB-57F 63-13502.
RB-57D	008	53-3971	4-6-56	6-15-56	Unk, MASDC c61, used as spares.
RB-57D	009	53-3972	3-5-56	6-1-56	4028 SRS, to RB-57F 63-13500.
RB-57D	010	53-3973	5-4-56	5-30-56	Unk, W-ADC, Cx 2-19-64 wings failed near Dayton, Ohio.

Type	Ship No.	AF Serial	Final Comp.	Flown Away	Remarks
RB-57D	011	53-3974	5-14-56	6-16-56	Unk, to RB-57F 63-13503.
RB-57D	012	53-3975	6-15-56	6-30-56	Unk, to RB-57F 63-13501.
RB-57D	013	53-3976	6-20-56	7-17-56	Unk.
RB-57D	001	53-3977	10-18-55	6-23-56	FAC 6-23 after mod, 4677 DSES c9-68, MASDC (A).
RB-57D	002	53-3978	11-18-55	2-8-56	Bailed J. Jones Cx 10-8-59.
RB-57D	003	53-3979	12-24-55	10-5-56	4028 SRS Cx Del Rio 10-2-58.
RB-57D	004	53-3980	12-28-55	6-23-56	FAC 6-23 after mod. (Martin entry), 4677 DSES c9-68, MASDC (A).
RB-57D	005	53-3981	1-13-56	5-9-56	Unk.
RB-57D	006	53-3982	2-24-56	4-29-56	Unk, 4677 DSES c9-68, Pima Air Museum Tucson, Arizona c93.
B-57E	336	55-4234	2-28-56	8-31-56	Bailed acc. 8-30-56.
B-57E	337	55-4235	3-23-56	6-9-56	7272 TTG, Eglin APG c9-68, MASDC c72 (A).
B-57E	338	55-4236	3-27-56	9-14-56	Unk.
B-57E	339	55-4237	4-18-56	9-14-56	Unk, Patricia Lynn 12-64, MASDC c72 (A).
B-57E	340	55-4238	4-24-56	10-1-56	Unk, Flagstaff Andrews 58-62 auto pilot, SEA Conv 65, 4677 DSES c9-68, Puerto Rico Educ/Scient Foundation (at Tucson, AZ c94).
B-57E	341	55-4239	4-27-56	9-14-56	Unk, MI ANG c9-68, 4713 DSES c12-72, MASDC c-76 (F).
B-57E	342	55-4240	5-1-56	9-14-56	Unk, 4713 DSES c9-68/c7-27-77, MASDC (F).
B-57E	343	55-4241	5-7-56	9-28-56	Unk, 4677 DSES c62, 4713 DSES c9-68/c7-27-77, MASDC (F).
B-57E	344	55-4242	5-9-56	10-5-56	Unk, 4677 DSES c9-68/c7-27-77, MASDC (F).
B-57E	345	55-4243	5-11-56	9-21-56	Unk, Star Flight Andrews 56-61, 58 WRS 61, Patricia Lynn 5-6-63, C-Loss 8-5-65 0 fat.
B-57E	346	55-4244	5-15-56	9-21-56	Unk, Eglin (AFSC) c9-68, Strategic Aerospace Museum Bellevue Neb c93.
B-57E	347	55-4245	5-17-56	8-30-56	Unk, Star Flight 58-62, 58 WRS 62, Patricia Lynn 5-6-63, MASDC c6-71 (A).
B-57E	348	55-4246	5-21-56	8-31-56	Unk.
B-57E	349	55-4247	5-23-56	9-14-56	5040 RES/5041 TOS Alaska, 21 Comp Wg AK, Cx 6-13-69 F-102.
B-57E	350	55-4248	5-28-56	9-14-56	AR ANG, SEA Conv 65, 8-BS Clark PI 3-17-66, MacDill 11-1-69, MASDC c72(A).
B-57E	351	55-4249	5-30-56	9-28-56	Unk, Patricia Lynn 12-64, MASDC 6-71 (D).
B-57E	352	55-4250	6-4-56	9-28-56	Unk, MASDC c72.
B-57E	353	55-4251	6-5-56	9-14-56	Unk, Star Flight Andrews 58-62, 58 WRS 62, SEA Conv 65, 4713 DSES c9-68, MASDC c2-79 (E).
B-57E	354	55-4252	6-7-56	9-28-56	Unk.
B-57E	355	55-4253	6-11-56	9-28-56	Unk, 4713 DSES c9-68-c7-77, AMARC c93.
B-57E	356	55-4254	6-13-56	10-3-56	5040 RES/5041 TOS Alaska, 21 Comp Wg AK, 4677 DSES 7-76, Cx 7-16-79 Salt Lake City en route MASDC, 0 fat.
B-57E	357	55-4255	6-15-56	10-3-56	Unk.
B-57E	358	55-4256	6-20-56	10-11-56	Unk.
B-57E	359	55-4257	6-22-56	2-5-57	Unk, Flagstaff, Convair c9-68, Patricia Lynn c10-68, 4677 DSES 6-71, MASDC c79 (E).
B-57E	360	55-4258	6-27-56	10-24-56	Edwards (AFSC) 60s, NB-57E, MASDC (A).
B-57E	361	55-4259	7-11-56	10-12-56	Unk, Star Flight Andrews 58-62, 58 WRS 61, SEA Conv 65, 4677 DSES c9-68, MASDC 7-79 (E).
B-57E	362	55-4260	7-9-56	10-12-56	Unk, 4713 DSES c9-68, MASDC 7-79.
B-57E	363	55-4261	7-17-56	10-13-56	Vincent AFB, Cx Yuma 6-3-57. 0-fat.
B-57E	364	55-4262	7-20-56	11-6-56	Edwards Test Plt Scl c9-68, MASDC c12-72 (A).
B-57E	365	55-4263	8-7-56	10-8-56	Unk, 4677 DSES, MASDC (E).
B-57E	366	55-4264	8-13-56	10-15-56	7407 CSW c8-64, Patrica Lynn 11-22-65, C-loss 25 Oct 68 0-fat.
B-57E	367	55-4265	8-17-56	10-17-56	7272 AGG c1959, SEA Conv 65, 8-BS Clark PI 3-18-66, (2 wheels up ldg 1-67, 6-67) MacDill 11-1-69, 4677 c12-72, MASDC c79 (F).
B-57E	368	55-4266	8-25-56	10-8-56	Unk, 4713 DSES c9-68, MASDC c79 (F).
B-57E	369	55-4267	10-19-56	11-29-56	Edwards (AFSC).
B-57E	370	55-4268	8-29-56	10-19-56	Unk, SEA Conv 65, Clark PI 3-1-66, RVN, C-Loss 6-13-66. 2 fat.
B-57E	371	55-4269	9-4-56	10-18-56	Unk, SEA Conv 65, Clark 3-14-66, RVN, LTV Compass Sight 7-15-68, Cx 10-14-71.
B-57E	372	55-4270	9-11-56	10-19-56	Unk, SEA Conv 65, 8-BS Clark PI 3-1-66, MacDill 11-1-69, 4677/17 DSES c12-72, MASDC c10-76.
B-57E	373	55-4271	9-12-56	10-23-56	Unk, Cx 1-20-58.
B-57E	374	55-4272	11-2-56	12-6-56	Edwards, Cx 6-17-58
B-57E	375	55-4273	9-18-56	10-29-56	Unk.
B-57E	376	55-4274	9-21-56	10-25-56	Unk, SEA Conv 65, Clark PI 1-17-66, RVN, LTV Compass Sight 7-15-68, 363 TRW c12-72, MASDC c10-76, Pima Air Mu c93.
B-57E	377	55-4275	9-25-56	11-19-56	6 TTS, 8 BS/TT, 6091 RS/556 RS c9-61, 18 TFW Kadena to 8-73, MASDC c10-76(E).
B-57E	378	55-4276	9-28-56	11-19-56	6 TTS, 8 BS/TT, 6091 RS/556 RS c9-61, 18 TFW Kadena to 8-73, 4677/17 DSES c7-26-77.

Type	Ship No.	AF Serial	Final Comp.	Flown Away	Remarks
B-57E	379	55-4277	10-3-56	11-29-56	6 TTS, 8 BS/TT, Cx 3-2-60.
B-57E	380	55-4278	10-8-56	12-1-56	6 TTS, 8 BS/TT, 4677/17 DSES c68-68, MASDC (F).
B-57E	381	55-4279	10-9-56	11-28-56	6 TTS, 8 BS/TT, 6091 RS/556 RS c9-61, 18 TFW Kadena to 8-73, 4677/ 17 DSES c7-26-77, Peterson AFB CO exhibit c93.
B-57E	382	55-4280	10-10-56	12-3-56	Unk, 4677 DSES c9-68, 17 DSES c79, MASDC (E).
B-57E	383	55-4281	10-16-56	12-1-56	6 TTS, 8 BS/TT, 6091 RS/556 RS c9-61, 18 TFW Kadena to 8-73, 4677/ 17 DSES c7-26-77, MASDC (E).
B-57E	384	55-4282	10-18-56	12-5-56	Unk, SEA Conv 65, 8-BS Clark 3-66, MacDill 11-1-69, MASDC c12-72 (A).
B-57E	385	55-4283	10-23-56	12-8-56	1 TTS, Cx Biggs AFB 1-18-57. 2 fat.
B-57E	386	55-4284	10-25-56	3-5-57	Unk, SEA Conv 65, Clark PI 3-1-66, RVN, C-Loss 12-13-68 w/C-123. 2 fat.
B-57E	387	55-4285	10-30-56	12-11-56	Unk, SEA Conv 65, 4677 DSES c9-68/c7-77, MASDC (F).
B-57E	388	55-4286	11-1-56	4-2-57	Unk, 4713 DSES c9-68, Cx Manitoba 2-7-69.
B-57E	389	55-4287	11-6-56	3-5-57	Unk, 4677 DSES c63, 4713 DSES c9-68, c79, MASDC (E).
B-57E	390	55-4288	11-8-56	3- 5-57	Unk, 4677 DSES c63, 4713 DSES c9-68, c79.
B-57E	391	55-4289	11-13-56	3-11-57	17 TTS, 4677 DSES c12-68, MASDC c12-72.
B-57E	392	55-4290	11-15-56	3-13-57	Unk, 4713 DSES c9-68, 4713 DSES c7-77, MASDC (E).
B-57E	393	55-4291	11-21-56	3-23-57	Unk, 4677 DSES c9-68.
B-57E	394	55-4292	11-26-56	3-23-57	Unk, 4677 DSES c9-68 to c7-77, MASDC (E).
B-57E	395	55-4293	11-30-56	3-20-57	Unk, 4677 DSES c9-68 to c7-77, Lowry AFB Museum Denver CO c93.
B-57E	396	55-4294	12-5-56	3-23-57	Unk, 4677 DSES c9-68, MASDC c10-76 (E).
B-57E	397	55-4295	12-10-56	4-2-57	Unk, 4677 DSES c9-68 to c7-77, MASDC (F).
B-57E	398	55-4296	12-11-56	4-2-57	Unk.
B-57E	399	55-4297	1-8-57	2-17-57	Unk.
B-57E	400	55-4298	1-14-57	2-11-57	Unk, 4713 DSES c9-68, 21 Comp Wg Alaska c12-72, 17 DSES c7-26-77, MASDC (F).
B-57E	401	55-4299	1-19-57	3-5-57	Unk.
B-57E	402	55-4300	2-4-57	4-2-57	Unk, 4713 DSES c9-68, c7-26-77,MASDC (F).
B-57E	403	55-4301	1-26-57	3-5-57	Unk.

MARTIN B-57 FINAL DISPOSITION SUMMARY

Although the negative aspects about the B-57 appear first when reviewing the above diagram, there are actually more positive things said when considering the airplane's service life of 27 years. Half of those built survived far beyond the years first intended for the B-57 and served those years quite well. Most of those were retired at AMARC with their service completed where they are eventually reclaimed as scrap and serviceable parts for other aircraft. This half also includes those resting in museums, all the RB-57Fs except two that crashed and the unknown amount that survived within the Pakistan Air Force. The other half of these airplanes were tragic losses, either due to crashes, combat losses including two disastrous ground explosions and those unknown losses within the Pakistan Air Force. Their many years as a service type aircraft speaks quite well for the total of 403 B-57s that were built and served these long years in so many different ways.

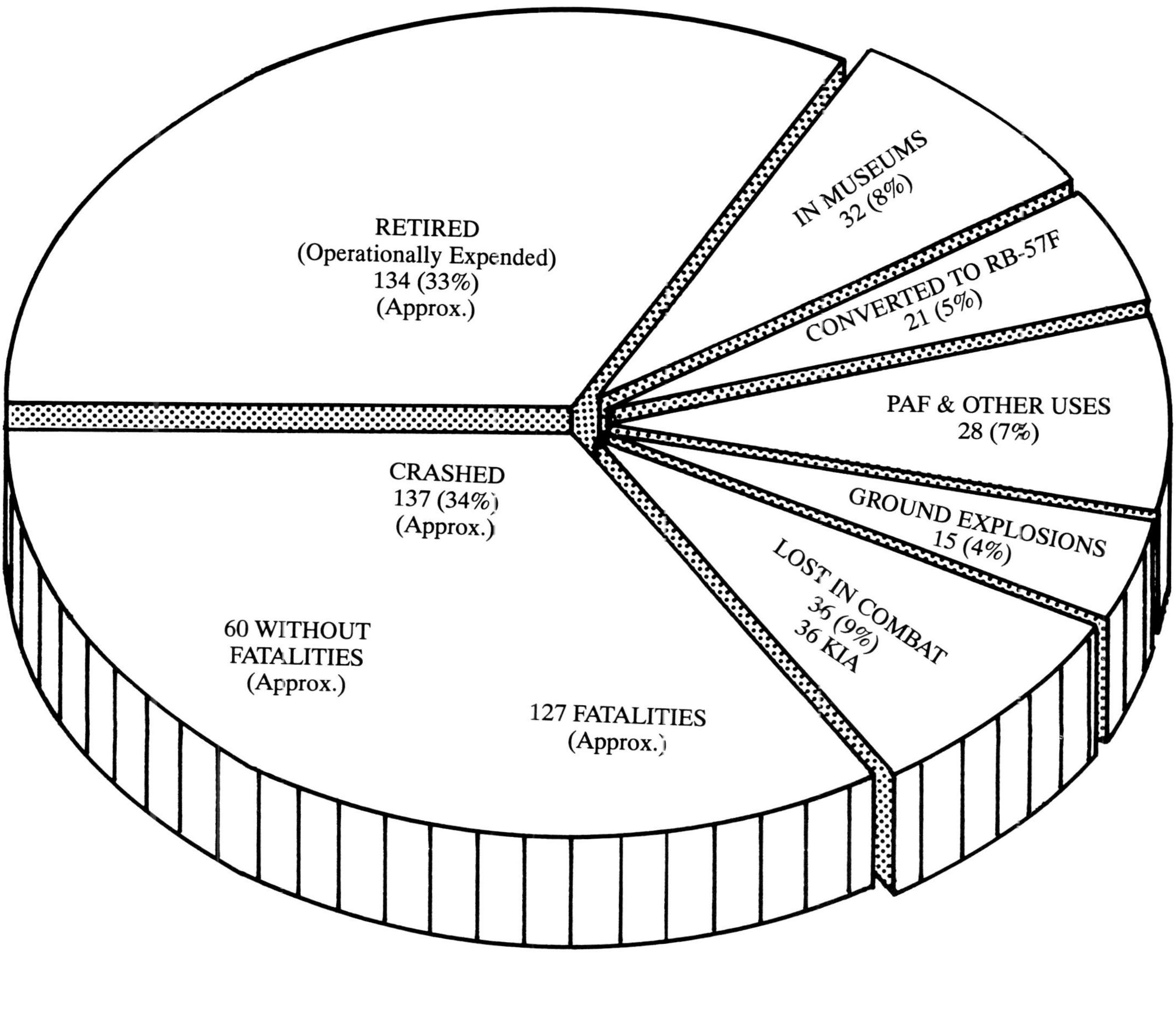

APPENDIX 14
MARTIN B-57 COMBAT ACTIVITY
8TH & 13TH BOMB SQUADRONS

Bases: BIEN HOA AB; DANANG AB; PAHN RANG AB

Period covered: April 1964 – December 1969

52-1512	8 Oct Dual flame-out at Clark. R.W.Clark/P.A. Viscasillas. 0 fat.
52-1523	21 Feb Combat loss near Hue. J.T.Stanley/H.E.Sjogren. 0 fat.
52-1532	Crash landed at Phan Rang. Combat loss. Zock/Miller. 0 fat.
52-1541	19 Sept Combat loss. Gould/Davis. 2 fat.
52-1555	1 Nov Destroyed by mortar attack at Bien Hoa AB.
52-1568	16 May Destroyed by accidental ground explosion at Bien Hoa AB.
53-3833	17 Apr Combat loss. L.R.Horacer/D.V.Harnage. 0 fat.
53-3836	4 Jun Training mission at Crow Valley range, Clark AB. K.Bratton/J.Bass. 2 fat.
53-3840	1 Dec To MacDill AFB for B-57G crew tng
53-3849	1 Dec To MacDill AFB for B-57G crew tng
53-3856	1 Dec To MacDill AFB for B-57G crew tng
53-3867	16 May Destroyed by accidental ground explosion at Bien Hoa AB.
53-3870	6 Aug Combat loss at Tan Son Nhut during deployment to Bien Hoa. F.C.Cutrer/L.L.Kaster. 2 fat.
53-3871	16 May Destroyed by accidental ground explosion at Bien Hoa AB.
53-3873	16 May Destroyed by accidental ground explosion at Bien Hoa AB.
53-3876	11 Jan Combat loss. Exploded on takeoff at DaNang. L.B.Smith/E.G.Tollett. 2 fat.
53-3877	20 Mar Transferred to B-57G mod.
53-3878	29 Mar Transferred to B-57G mod.
53-3879	1 Oct To MASDC storage.
53-3880	6 Apr Combat loss over Laos. J.M.Lewis/A.D.Baker. 2 fat.
53-3882	8 Jun Combat loss. Ejected over Can Tho due to hit in wing and fire. G.Nelson/J.Carnes. 0 fat.
53-3884	6 Aug Collision on landing at Bien Hoa AB with 877. 0 fat.
53-3886	1 Jun Transferred to B-57G mod.
53-3888	6 Oct Combat loss. G.Rippey/L.F.Makowski 0/1 fat.
53-3889	1 Jun Transferred to B-57G mod.
53-3890	11 Mar Combat loss. Dropped 500 GPs low instead of napalm and caught in bomb blast near Kontum. W.C.Mattis/R.D.Smith. 2 fat.
53-3892	1 Nov Destroyed by mortar attack at Bien Hoa AB.
53-3893	16 May Destroyed by accidental ground explosion at Bien Hoa AB.
53-3894	1 Nov Destroyed by mortar attack at Bien Hoa AB.
53-3895	29 Jun Combat loss. Wing fire and explosion near Tan Son Nhut. S.P.Chambers/R.G. Landringham. 2 fat.
53-3898	1 Jun Transferred to B-57G mod.
53-3903	16 Jan Combat loss. Into water off DaNang. E.J.Cook/H.M.Davidson. 0 fat.
53-3904	16 May Destroyed by accidental ground explosion at Bien Hoa AB.
53-3905	15 Jul Transferred to B-57G mod.
53-3906	1 Jun Transferred to B-57G mod.
53-3908	9 Jun Combat loss. 2 fat.
53-3910	21 Jun Combat loss during night flare mission over N. Vietnam. C.K.Lovelace/W.E.Cordero. 2 fat.
53-3913	16 May Destroyed by accidental ground explosion at Bien Hoa AB.
53-3914	1 Nov Destroyed by mortar attack at Bien Hoa AB.

1964 · 1965 · 1966 · 1967 · 1968 · 1969 (monthly timeline, April 1964 – December 1969)

BIEN HOA AB · DANANG AB · PAHN RANG AB

KENTUCKY ANG · NEVADA ANG · KENTUCKY & NEVADA ANG

53-3915	16 May Destroyed by accidental ground explosion at Bien Hoa AB.
53-3919	6 Aug Combat loss. Crashed in Nha Trang city. L.R.Horacek/R.Johnson. 0 fat.
53-3924	1 Nov Destroyed by mortar attack on Bien Hoa AB
53-3925	14 Apr Operational loss, mid-air while in formation aerobatics with 926 at DaNang. R.P.Bateman/R.H.McCaw. 0 fat.
53-3926	14 Apr Operational loss, mid-air while in formation aerobatics with 925 at DaNang. J.T.Greshel/V.G.Wittkopp. 0 fat.
53-3930	16 May Destroyed by accidental ground explosion at Bien Hoa AB.
53-3931	1 Jun Transferred to B-57G mod.
53-3937	16 May Destroyed by accidental ground explosion at Bien Hoa AB.
52-1507	From KY ANG 1 Jan/1 Oct to MASDC storage.
52-1508	From KY ANG 1 Jan/27 Mar Combat loss near Phu Cat. 0 fat.
52-1519	From KY ANG 1 Jan/1 Oct To MASDC storage.
52-1551	From KY ANG 1 Jan/15 Oct To MASDC storage.
52-1561	From KY ANG 1 Jan/13 Jan Combat loss. Eaton/Getchel. 2 fat.
52-1575	From KY ANG 1 Jan /10 Feb Combat loss. Crew ejected, no recovery. R.P.Hunter/E.P.Kiefel. 2 fat.
52-1589	From KY ANG 1 Jan/30 Jun Combat loss. Aborted takeoff, ensuing damage exploded bombs. R.N.Daeke/L.K.Rygg. 0 fat.
53-3828	From KY ANG 1 Jan/1 Oct To MASDC storage.
53-3940	From KY ANG 1 Jan/1 Nov To MacDill for B-57G tng.
52-1592	From NV ANG 20 Apr/25 Mar Lost control from single-eng landing. R.W.Hopper/D.L.McHogo. 2 fat.
53-3920	From NV ANG 20 Apr/20 Oct Combat loss. G.T.Hamilton/H.E.Holzapple. 0 fat.
52-1518	NV/KY ANG 23 May/22 May To *Tropic Moon* Project
52-1530	NV/KY ANG 23 May/7 Apr Combat loss. R.Barnett/T.Walker. 2 fat.
52-1544	NV/KY ANG 23 May/6 Sep Engine fire indication. W.C.Hamann/R.J.Lane. Combat loss. 0 fat.
52-1545	NV/KY ANG 23 May/1 Oct To MASDC storage.
52-1565	NV/KY ANG 23 May/14 Dec Combat loss near Pleiku. R.J.Moroney/J.Joiner. 1/0 fat.
52-1567	NV/KY ANG 23 May/15 Mar Dual flame out near Pleiku. Tiddy/M.De Sousa. 0 fat.
52-1570	NV/KY ANG 23 May/5 Oct Combat loss at Phan Rang. Miller/R.Dyer. 0 fat.
53-3860	NV/KY ANG 23 May/4 Jun To *Tropic Moon* Project.
53-3929	NV/KY ANG 23 May/1 Jun Transferred to B-57G mod.
52-1498	Arrival unknown/22 Sep Combat loss N-VN D.Klien/Erickson. 0 fat.
52-1499	Mar 21/1 Oct To MASDC storage.
52-1510	11 Dec/14 Sep Combat loss. Overly/nav. 2 fat.
52-1516	1 Aug/1 Oct To MASDC storage.
52-1550	20 Dec/6 Mar Belly landed at DaNang. R.Barnett. 0 fat.
52-1557	Arrival unknown/21 Jan Combat loss. G.Cooper/G.Low. 2 fat.
52-1571	21 Jan/1 Oct To MASDC storage.
52-1578	Arrival unknown/1 Jun Transferred to B-57G mod.
52-1580	Arrival unknown/19 Jun *Tropic Moon* Proj. and B-57G
52-1582	Arrival unknown/1 Jun Transferred to B-57G mod.
52-1586	30 Jul/4 Apr Combat loss. Zock/nav. 0 fat.
52-1588	22 May/1 Jun Transferred to B-57G mod.
52-1590	11 Jan/8 Dec Combat loss. L.Wilkinson/nav. 0/1 fat.
53-3827	10 Dec/1 Dec Trans to MacDill AFB for B-57G tng.
53-3838	22 Dec/8 Jan Tng landing accident at Cubi Point, PI. G.A.Blake/VNAF plt. 0 fat.
53-3859	Arrival unknown/22 Apr Combat loss. J.E.Oxley/C.W. Hohman. 2 fat.
53-3862	Arrival unknown/3 Jun Combat loss N-VN. T.Springston/nav. 2 fat.
53-3865	Mar 15/1 Jun Transferred to B-57G mod.
53-3928	17 Apr/15 Jul Transferred to B-57G mod.
55-4248	17 Mar/1 Oct Trans to MacDill AFB.
55-4265	18 Mar/1 Oct Trans to MacDill AFB.
55-4268	1 Mar/13 Jun Combat loss C.W.Burkhart/E.O.Kerr. 2 fat.

	1964	1965	1966	1967	1968	1969
	April–December	January–December	January–December	January–December	January–December	January–December
	BIEN HOA AB (August 1964 – June 1965)	DANANG AB (July 1965 – March 1966)	PAHN RANG AB (April 1966 onward)			

Aircraft	Notes
55-4269	14 Mar/15 Jul Transferred to LTV, *Compass Sight*.
55-4270	1 Mar/1 Oct Trans to MacDill AFB.
55-4274	17 Jan/15 Jul Transferred to LTV, *Compass Sight*.
55-4282	Arrival unknown/Trans to MacDill AFB.
55-4284	1 Mar/13 Dec Combat loss. Dugan/F.J.McGouldrick. 2 fat.

APPENDIX 15
GLOSSARY

AA	Antiaircraft.
AAA	Antiaircraft Artillery.
AB	Air Base, when on non-US territory.
ADC	Air Defense Command.
(A)DC	(Air) Development Center.
ADVON	Advanced echelon.
AEC	Atomic Energy Commission.
AFB	Air Force Base, usually that of USAF.
AGG	Aircraft Gunnery Group.
AGL	Above ground level.
AM	Alminum Matting, often used for temperary hard surfaces, i.e. runways, hardstnds.
AMARC	Aerospace Maintenance and Regeneration Center.
AMC	Air Materiel Command.
ANG	Air National Guard.
APG	Air Proving Ground (at Eglin)
API	Armor-Piercing-Incendiary, type of ammunition.
ARVN	Army of the Republic of Vietnam.
ASD	Air Systems Development Command.
ATS	Aerial Tracking Squadron.
BDA	Bomb damage assessment.
BG	Bombardment Group.
BOQ	Bachelor officers quarters.
BW	Bombardment Wing.
cal.	A measurement system for diameters of projectiles.
CCTS	Combat Crew Training Squadron
CCTW	Combat Crew Training Wing
cg	Center of gravity.
CIA	Central Intelligence Agency.
CIDG	Civilian Irregular Defense Group.
CO	Commanding officer, of a specific unit.
CSG	Combat Support Group.
DASC	Direct Air Support Center.
deg	Degree.
Det.	Detachment, usually part of a squadron.
DMZ	Demilitarized Zone
DOE	Department of Energy.
DOOM	Da Nang Officers Open Mess.
DSEG	Defence Systems Evaluation Group.
DSES	Defense Systems Evaluation Squadron.
EAS	Earth Survey Aircraft.
ECM	Electronics Counter Measure.
ERDA	Energy Research & Development Agency.
ERSP	Earth Remote Sensing Platform.
ERTS	Earth Resources Technology Satellite.
ESCAPAC	Escape packet, a type of ejections seat with survival equipment.
EWO	Electronics Warfare Officer.
FAA	Federal Aviation Administration.
FAC	Forward Air Controller.
FIS	Fighter Interceptor Squadron.
FY	Fiscal Year.
G	Gravity, to be subject to G-forces.
gal	Gallon measurement, U.S. gal unless specified.
GCA	Ground Controlled Approach, for radar guidance.
GCI	Ground Controlled Intercept.
GD/FW	General Dynamics, Fort Worth Division.
GP	General Purpose, a type of demolition bomb.
HE	High Explosive, filler for bomb or shell.
IAF	Indian Air Force.
IAS	Indicated airspeed.
IFF	Identification, friend or foe, a transponder.
IR	Infrared, refering primarily to photography.
kt	Knot; a nautical mile per hour.
Lacie	Large area crop inventory experiment.
LASL	Los Alamos Scientific Laboratory.
lb	Pound measurement abbreviation.
Mach.	Speed measurement in terms of sound in the medium in which it moves. Speed of sound is Mach 1.
MACV	U.S. Military Assistance Command, Vietnam.
MDAP	Mutual Defense Assistance Pact.
MIA	Missing in action.
MIL SPEC	Military Specifications.
min	Minutes
Mk	Mark, meaning a type followed by a number.
NATO	North Atlantic Treaty Organization
NAS	Naval Air Station.
NASA	National Aeronautics and Space Agency.
NCAR	National Center for Atmospheric Research.
nm	Nautical mile.
NORAD	North American Defense Command
OG	Operations Group.
OS	Operations Squadron.
PACAF	Pacific Air Force, part of USAF.
PAF	Pakistan Air Force.
Panama	Call sign for the control and reporting center at Da Nang Air Base.
PBR	Patrol boat, rescue.
PCS	Permanent Change of Station.
PSP	Pierced Steel Planing, a matting often on which to park aircraft.
RAAF	Royal Australian Air Force.
RADC	Rome Air Development Center (AFSC)
RAPCON	Radar approach control center.
RES	Radar Evaluation Squadron.
RESCAP	Rescue Air Patrol.
RS	Reconnaissance Squadron
RVN	Republic of Vietnam (South Vietnam)
SAM	Surface-to-Air Missile, Soviet built.
SEATO	Southeast Asia Treaty Organization.
SHORAN	Short range navigation, a radar ranging system.
SIOP	Single Integrated Operations Plan for striking targets in the event of an atomic war.
SRS	Strategic Reconnaissance Squadron.
SRW	Strategic Reconnaissance Wing.
TAC	Tactical Air Command.
TACAN	Tactical Air Navigation, a instrumented system.
TAS	True Airspeed.
TBG	Tactical Bombardment Group (later Wing).
TBS	Tactical Bomgardment Squadron.
TDY	Temporary Duty, away from home station.
TEWS	Tactical Electronics Warfare Squadron.
TFR	Terrain following radar.
TFW	Tactical Fighter Wing.
TG	Tactical Group.
TOS	Tactical Operations Squadron.
TRS	Tactical Reconnaissance Squadron.
TRG	Tactical Reconnaissance Group.
TRW	Tactical Reconnaissance Wing.
TSN	Tan Son Nhut, station identifier for Saigon Airport.
TTF	Tow Target Flight.
TTS	Tow Target Squadron.
USAFE	United States Air Force, Europe.
Viet Cong	Viet Nam Cong San: Vietnamese Communist, usually South Vietnamese Communists.
VNAF	Vietnamese Air Force.
VOR	VHF Omni-Directional Range, for navigation.
WADC	Wright Air Development Center.
Wingman	Pilot who flies at the side or rear of his element leader.
WRG	Weather Reconnaissance Group
WRS	Weather Reconnaissance Squadron
WRW	Weather Reconnaissance Wing
WRS	Weather Reconnaissance Squadron.
Wx	Weather.

INDEX

People

Aircraft

Air Bases

Numbered Units

Tactical Bombardment Units

Reconnaissance Units

Tow Target and DSES Units

Air National Guard Units

Other

Operations and Projects

Southeast Asian Locations
(Also see Air Bases)

Companies and Corporations

General

V
033856
USAF
BA-856
U.S. AIR FO